Generis
PUBLISHING

TESTING OF COMPOSITE MATERIALS USED IN WIND TURBINE BLADES

Ciprian-Ionuț Morăraș
Viorel Goanță

Title: **TESTING OF COMPOSITE MATERIALS USED IN WIND TURBINE BLADES**

ISBN: 979-8-89966-090-0

Author: Ciprian-Ionuț Morăraș, Viorel Goanță

Cover image: Dr. Ciprian Morăraș and Prof. Dr. Viorel Goanță

Publisher: Generis Publishing
Online orders: www.generis-publishing.com
Contact email: info@generis-publishing.com

Ciprian-Ionuț Morăraș

Viorel Goanță

TESTING OF COMPOSITE MATERIALS USED IN WIND TURBINE BLADES

Contents

Preface

The subject approached within this book, which is of obvious theoretical and practical importance, has been a preoccupation of the authors in recent years. By analyzing the current state of research in the field of regenerative energy and wind turbine blades, some shortcomings have been found in testing blades at compound stresses. The paper presents a system for laboratory testing of small-scale blades to reproduce as closely as possible the stresses during operation.

In recent years, wind power generation has grown exponentially, both internationally and nationally. As a result of research into wind turbines, the cost of generating electricity from wind has become increasingly competitive. Wind energy is gaining more and more ground in the energy industry as a renewable, carbon-neutral source. A major trend is the development of larger wind turbines, which cost less to power. However, the failure of wind turbines is still a serious problem, especially as blade damage can cause major structural failures. In the overall wind turbine system, the blade is the least reliable element. This justifies the importance of research into increasing the reliability of wind turbine blades.

The book is in line with international and national trends on increasing the performance of wind turbines and the share of renewable energy to reduce global warming. Due to the increase in rotor diameter, the replacement of metallic materials by composite materials such as GFRP or CFRP has become a current priority in wind turbine blade construction. Thus, the topic fits into modern, current and highly prospective directions for today's world priorities. The applicability of the research conducted in the book, on increasing wind turbine blade reliability, follows and tests the main component in a wind turbine assembly.

Based on the realization that blades are the least reliable element of the wind turbine assembly, the authors aim to develop research on both the composite material of which the blade is made and the laboratory testing of small blades, which reproduce as closely as possible the stresses that occur during operation. The following objectives were achieved: complex testing of glass fiber reinforced composite (GFRP) materials under different humidity conditions; identification of the failure mechanisms of the composites; finite element analysis of a torsional buckled blade model; testing the torsional buckled blade model, acquisition and processing of experimental data; comparison of the results predicted with FEA with those obtained experimentally;

identification of possible failure modes of the blade model; formulation of conclusions and recommendations.

Thus, the authors make contributions to increase the reliability of wind turbine blades, which leads to the improvement of operational safety of wind turbines and the development of renewable energy.

Summary

This work entitled *„**Testing of composite materials used in wind turbine blades**"* is the result of extensive research within the "Gheorghe Asachi" Technical University of Iaşi, Faculty of Mechanics, Department of Mechanical, Mechatronics and Robotics Engineering.

Chapter 1 presents the current state of research, which includes a synthesis study on the development of wind energy in the European Union and Romania. It also describes the constructive elements of a wind turbine, the element of interest being the turbine blade. Due to its complex structure and mode of operation, a synthesis of risk factors and defects that may occur during operation is carried out. It is important to study the material from which the turbine blade is made through various mechanical tests (tensile, compression, bending, shear).

Chapter 2 presents experimental research on the characteristics of the GFRP glass fiber reinforced composite material with fiber orientation at [0°/90°] and [-45°/0°/+45°/90°]. For the tensile tests, five specimens were used, cut longitudinally (relative to the plate format), transversely and diagonally at 45° from the plate with fiber orientation at [0°/90°]. Five longitudinal and transverse specimens were made from the board with fibers oriented at [-45°/0°/+45°/90°]. Static tests were performed in accordance with ASTM D3039 and ASTM D3518 standards. In this chapter, the resistive electrical strain gauge method was used to determine the material constants. Another test used for the composite material specimens was the compression test on the plate with the fiber orientation at [0°/90°]. Pure shear tests were also performed on flat specimens, cut from the two composite plates reinforced with fiberglass fabric. Knowing that wind turbine blades operate in unfavorable conditions (bad weather, low temperatures), water immersion tests were also necessary for 7, 14, 21 and 31 days. In addition, as the turbine blade can operate at very low temperatures, the immersion of a specimen in liquid nitrogen at a temperature of -70°C for four days was also investigated. Electron microscopy analyses were performed using the SEM Quanta 2003D electron microscope, on the samples with the highest resistances. The aim was to highlight the clear aspects of fiber breakage and characteristic areas.

Chapter 3 contains tests performed on CT (compact tension) specimens made of GFRP glass fiber reinforced composite with fiber orientation at [0°/90°] and [-45°/0°/+45°/90°]. The determination of the crack propagation velocity was followed and graphs were drawn only for the most representative samples in terms of the results obtained. After the pre-cracking tests, defectoscopic determinations were used to visualize and measure the crack initiated by fatigue propagation. The determination of cracks in CT specimens was performed using penetrant liquids, respecting all stages. The last subchapter presents the determination of the fracture toughness.

Authors, 2.07.2025

CHAPTER 1. STATE OF THE ART IN WIND ENERGY CONVERSION SYSTEMS

1.1 Renewable energy and sustainable development of society

1.1.1 Energy and environmental pollution

With the onset and development of industrialization, energy consumption, both globally and locally, has seen an accelerated increase, for various reasons. Although there is no known completely "clean" energy source, however, traditional energy sources are much more polluting, contributing to increased greenhouse gas emissions, which have dramatic consequences on global warming. Developing renewable energy will make a considerable contribution to mitigating climate change. Fortunately, governments have realized the importance of protecting the environment and are taking action to reduce the long-term impact of the energy industry and beyond. Sustained strategies are being developed to increase energy use efficiency and implement energy efficiency programs. Economic development depends, to a large extent, on the ability to secure energy needs at a reasonable price. Renewable energy resources, as well as technologies with the least possible impact on the environment, are a priority for today and the future.

For the first time, in 2020, renewable energy overtook fossil fuels as the EU's main source of energy, with 38% of electricity produced in the EU coming from renewable sources, 37% from fossil fuels and 25% from nuclear power. The European Climate Act requires greenhouse gas emissions to be reduced by 55% by 2030, with climate neutrality to be achieved by 2050 (Raport, 2021).

Romania, as an EU country, will have to implement these requirements. According to ANRE's Annual Activity Report 2023, in the period 2022-2031, renewable sources will register a total capacity increase of 6615 MW (of which 6127 MW wind + solar), the total capacity of renewable energy sources installed in the national electro-energy system will exceed 17 GW (including hydro).

The evolution of the total net generation capacity available in the national power system in the target years 2022, 2026 and 2031, depicted in Figure 1.1, illustrates the reduction of the share of coal (in particular lignite) in the generation mix in 2026 and its future replacement by natural gas, renewables and nuclear, in line with the EU commitments to reduce greenhouse gas emissions.

Romania has a high wind potential, but it is still insufficiently exploited. Thus, in the high mountain area wind resources at 50m height exceed $1800W/m^2$, in the coastal area over $700W/m^2$, and in the hills and foothills up to 250 W/m^2 (wind energy). In order to comply with European legislation, Romania will have to develop renewable energy sources at an accelerated pace, first and foremost by capitalizing on its wind potential. To this end, new large-scale wind turbine parks will be built, but the number of medium and small wind turbines intended to achieve energy independence for local consumers will also increase.

The latter could, at certain times, supply electricity into the national grid, becoming energy consumers and producers ("prosumers").

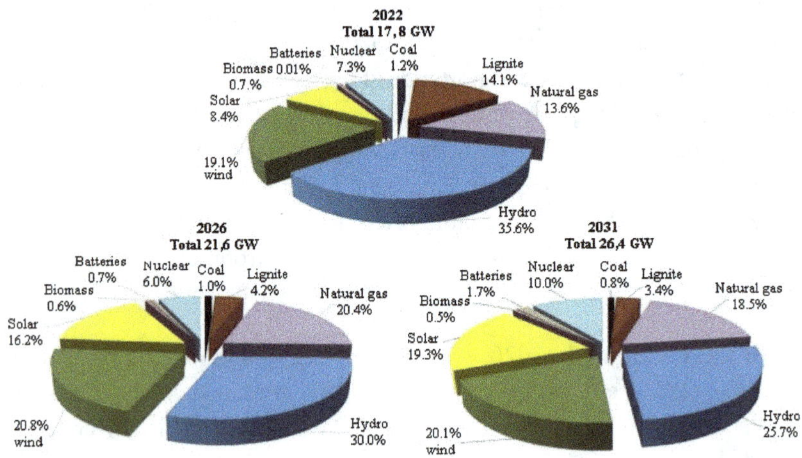

Fig. 1.1 Electricity structure in Romania in 2022 (Annual report on ANRE 2023)

1.1.2 Developing wind energy production

Globally, the installed capacity of wind turbines has grown rapidly over the last 20 years, exceeding 790MW in 2020, Figure 1.2. This growth trend will accelerate in the coming period.

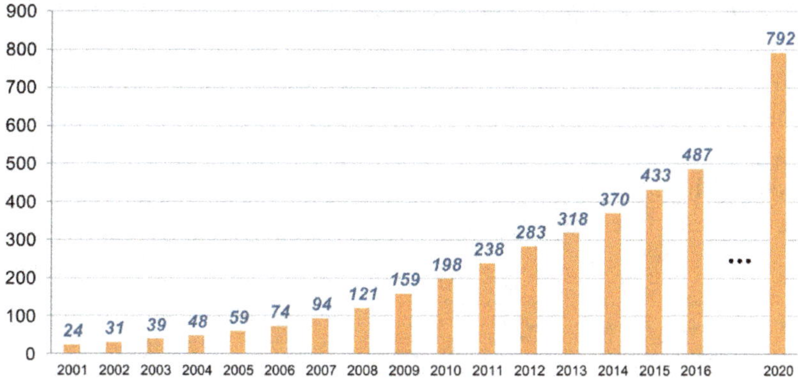

Fig. 1.2 Increasing wind power capacity world-wide [GW] (F. Blaabjerg, 2017)

The construction of wind turbines with the highest possible power rating is the main way of lowering the price of wind-generated electricity. This means making the rotors as large as possible. The largest turbines have a rated power of over 10MW and a rotor diameter of about 190m, Figure 1.3. For a 190m diameter rotor the blade is 83m long. At nominal rotor speed of 13 rpm, the blade tip speed is about 300km/h.

The realization of large rotors would not be possible without the use of composite materials for the construction of the blades. The cost of the three blades represents about 15-20% of the total cost of the wind turbine. Replacing one blade can cost up to $200,000, and one day of turbine downtime results in losses of $800-1600 (Shohag, M.A.S and others, 2017; Mishnaevsky, 2020; Burton T and others, 2001).

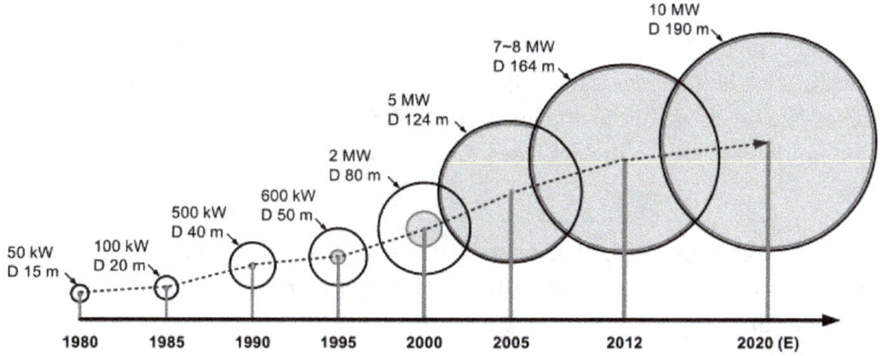

Fig. 1.3 Increasing the rated power and propeller diameter of wind turbines (Yang, 2013)

1.2 The structure of wind turbine blades

Wind turbine blades have a complex structure, with the typical cross-section being that of a thin-walled multi-cone multi-cone profile. Several materials are used to manufacture the blades, which are blended together, Figure 1.4: metal alloys, composites (fiber-reinforced, fabric-reinforced, sandwich), protective materials (primer, UV protection gels, paint, etc.).

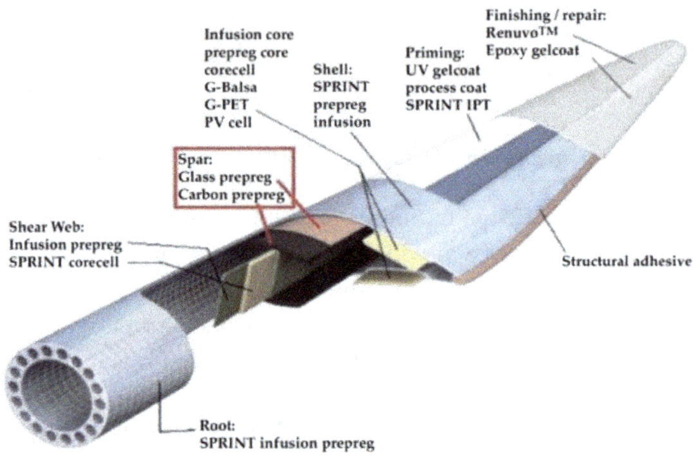

Fig. 1.4 Materials used in the construction of wind turbine blades (Steigmann et.al., 2016)

1.3 Breaking the blades

With increasing length, the mechanical stresses in the blade increase and the probability of failure increases. The blade is the least reliable element of the wind turbine. The breakage of one blade causes the rotor to become unbalanced and usually results in damage to the other blades due to impact with fragments from the broken blade.

Between 2015-2020, there were 26 major wind turbine accidents worldwide. Out of these, 17 (65.38%) were due to blade failure, Figure1.5. In addition to breakage, blades can also fail due to other causes such as: environmental action, composite material defects, etc.

Wind turbine blades may have defects introduced during the manufacturing process, or, they may occur as a result of operation. Large manufacturing defects are rectified at the factory. Small defects may increase their original size due to operational stresses: variable stresses, shocks, compound stresses, environmental factors. The most common manufacturing defects are:

- exfoliation of the protective layer;
- inadequate adhesive joints;
- delaminations in the composite material;
- failures of individual laminae;
- delaminations and cracks in the protective layer;
- delaminations between the fabric and the matrix.

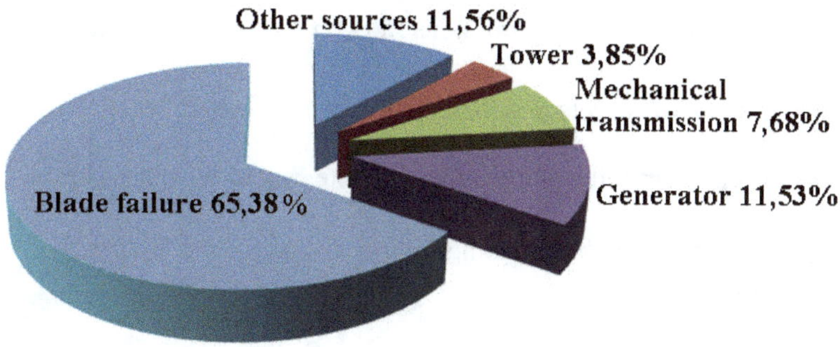

Fig. 1.5 Major wind turbine failures, occurring between 2015 and 2020, and their causes (Wind turbine accident and incident compilation)

For these reasons, the increase in wind turbine reliability is predominantly determined by the increase in blade reliability. This justifies focusing research on the development of blades with the lowest possible risk of failure.

1.3.1 Wind turbine blade defects

Damage to wind turbine blades can occur during the manufacturing process, during transport, assembly or during operation. Large faults in the technological process, which may be capable of compromising the proper functioning of the blade, are rectified by the manufacturing company itself. Small defects can lead to micro-cracks, which increase in size as a result of stresses during operation (compound stresses, impact, environmental factors). These cracks can reach critical size over time, and for this reason, it is recommended to periodically inspect the blades and/or monitor them during operation using sensors (Xiaowen Song et al., 2022).

Nowadays, the manufacturing process of blades is partially automated and thus many manufacturing defects have been eliminated. There are still, however, some critical areas of the blade that are not accessible to standard non-destructive testing (NDT) techniques:

- strain gauge bonded to the surface of the blade or incorporated into the composite material during blade manufacture;
- optical fibers integrated in the blade material (method Fiber Bragg Gratings -FBG);
- ultrasonic inspection;
- thermography;
- detection of acoustic emissions (50 kHz to 1 MHz) occurring during blade operation due to fiber fracture or matrix cracking;
- *Machine Vision* techniques such as Digital Image Correlation (DIC).

The most frequent failures that affect the integrity of the blades are:

- damage caused by lightning;
- fatigue collapses;
- leading edge erosion;
- damage due to ice deposition;
- moisture damage, etc.

Figures 1.6 and 1.7 show the critical areas of blade damage due to mechanical loading.

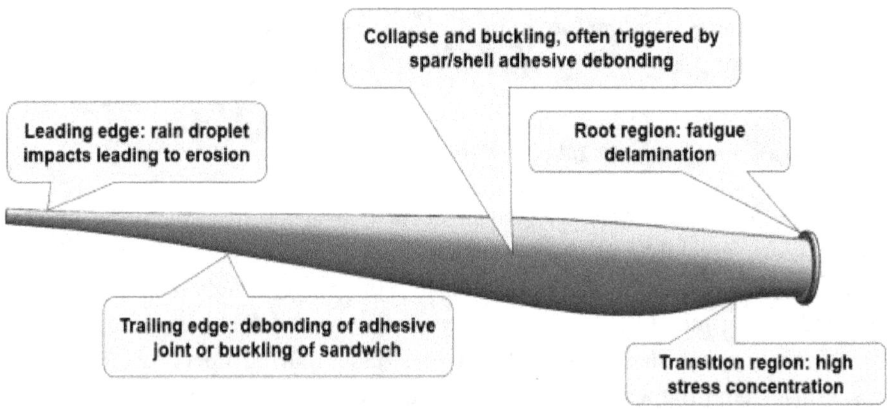

Fig. 1.6 Critical areas of damage caused by the mechanical loading of the blade (Abdel-Hamid et.al., 2023)

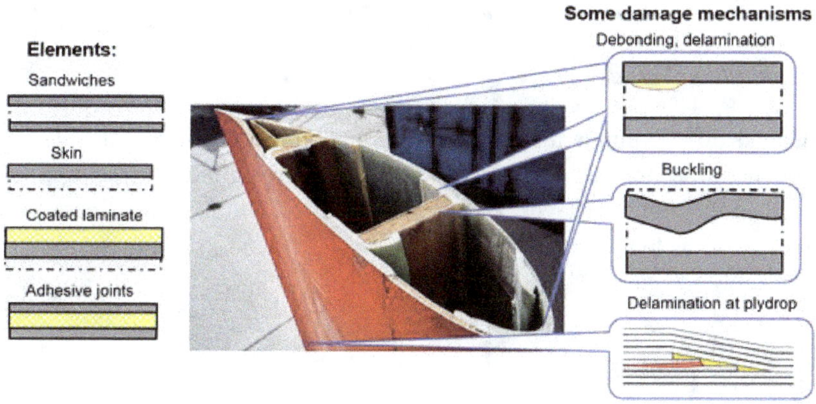

Fig. 1.7 Typical delamination and debonding damage in different areas of the blade (Mishnaevsky, 2022)

Figure 1.8 shows the failure mechanism of the joint between the box-section stringer and the shell, which includes delamination, propagation and fracture.

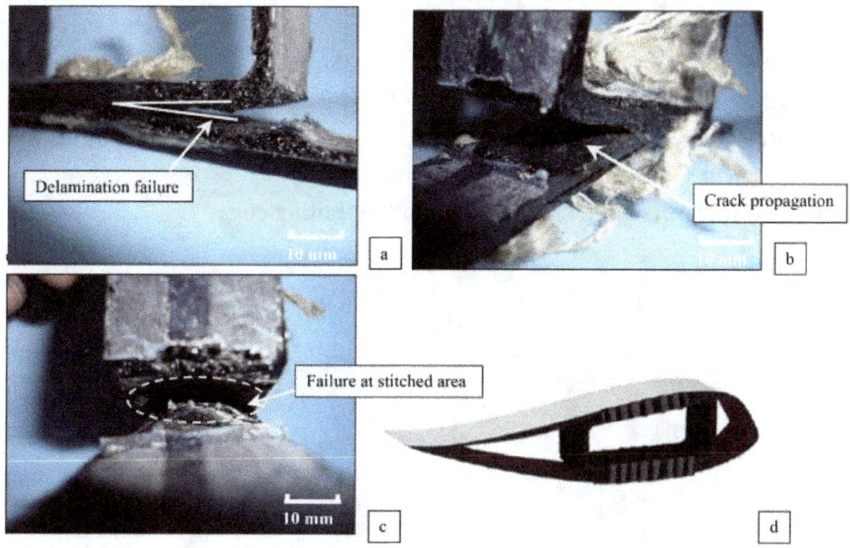

Fig. 1.8 The failure mechanism of the joint between the boxed stringer and the shell (Al-Khudairi et.al., 2015)

Cracks in wind turbine blades

Longitudinal cracks

Cracks 50-400 mm in length were observed both at the base of the blades and in the aerodynamic zone. The largest longitudinal crack reported was 600 mm long and developed in the adhesive layer. Debonding up to 1.35 m long was recorded on the trailing edge of the blades.

Transversal cracks

Transverse cracks with lengths up to 40 mm were observed mainly on the trailing edge. Transverse cracks were classified as follows, Figure 1.9:

- Simple cracks, which occur on only one of the two halves of the blade and stop at the blade edge;
- Cracks on the circumference, extending on both the intrados and extrados of the blade;
- Short cracks occurring at the base, 15-25 mm in length, singly or in groups of up to eight cracks, with a pitch of about 20-60 mm;
- Repeated simple cracks, usually no deeper than the thickness of the overlay, occurring due to vibration from wind gusts;
- 45° transverse cracks appearing on the trailing edge.

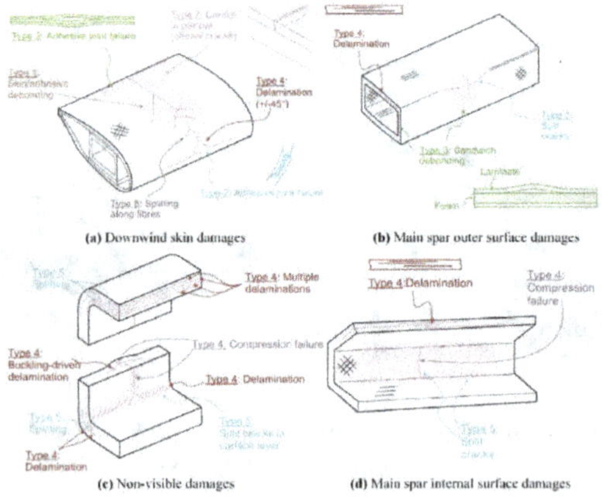

Fig. 1.9 Failure modes observed in a wind turbine blade tested to failure (Soerensen, 2000; Abdul Ghani Olabi, et.al., 2021)

Erosions to the leading and trailing edge

The inspection of 36 wind turbine blades, in service for 17-22 years, revealed the following major of trailing edge defects, Table 1.1, (Ataya S, 2013):

- Longitudinal cracks between bonded materials at a distance of about 0.73m from the rotor radius;
- Transverse cracks, which occur mainly on the trailing edge, in the area most subject to fatigue stress;
- Cuts and crushed.

Table 1.1 Number of defects found on the trailing edge of the blade (Ataya S, 2013)

Discontinuities	Type of turbine		Total
	ETB 300 kW	ETB 100 kW	
Longitudinal cracks	79	57	136
Transversal cracks	14	59	73
Cut and crushed at the edges	43	40	83
Total	136	156	292

Cuts, crushing and erosion were observed in the area of the leading and trailing edges Figure 1.10, (Yajie Yua et al., 2020) which may occur due to interaction with other solid bodies. On a 300kW turbine blade, the presence of cuts on the trailing edge was reported, having a length of ca. 34 mm, about 3.4 m from the blade tip.

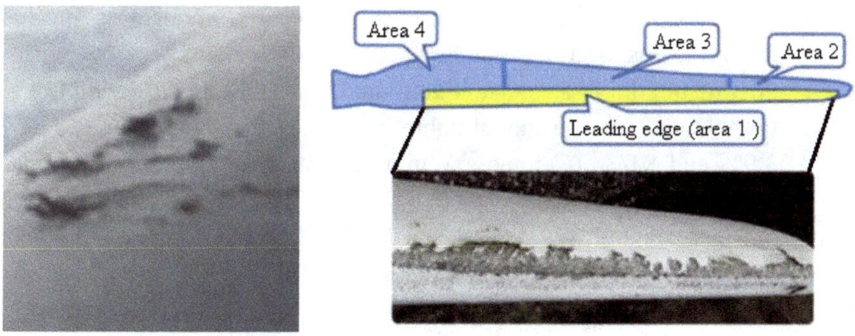

a) b)

Fig. 1.10 Damage to the trailing edge: cuts and crushing (a); erosion(b), (Yajie Yua et.al., 2020; X. Chen, et.al.,2014)

The number of blades broken was found to be proportional to their size. Figure 1.11 shows the breakage of a blade.

a) **b)**

Fig. 1.11 A damaged wind turbine blade (a) and cracks in the blade (b) (Damage blade)

The leading edge takes the most impact from wind-driven particles and bodies. These wear the blade surface. Blade erosion can be caused by wind-borne sand and pebbles, hail, rain, snowfall, insects and sea water, etc. Protective coatings are used to minimise blade erosion. On unprotected blades from the factory, signs of erosion appear in less than 3 years. Defects in the protective coating lead to erosion of the leading edge and even of the entire blade.

Other factors of damage to the blades

The most common damage to blades is caused by:

- 40% are mechanical failures, which occur in blades, mechanical transmissions, bearings and the nacelle rotation system;
- 20% of all failures are due to lightning, Figure 1.12. In the period from 1991 to 1998, the average annual lightning outages in Europe averaged between 3.9% and 8.0%, (Caithness Windfarm Information Forum. Summary of Wind Turbine Accident Data to 31 March 2021). Between 2002 and 2006, in Japan, lightning caused about 10-20% of all failures (Ideno, 2006);
- 9% of all damage is due to fires, Figure 1.13.

Fig. 1.12 Lightning damage to the blade tip <u>Wind power</u>

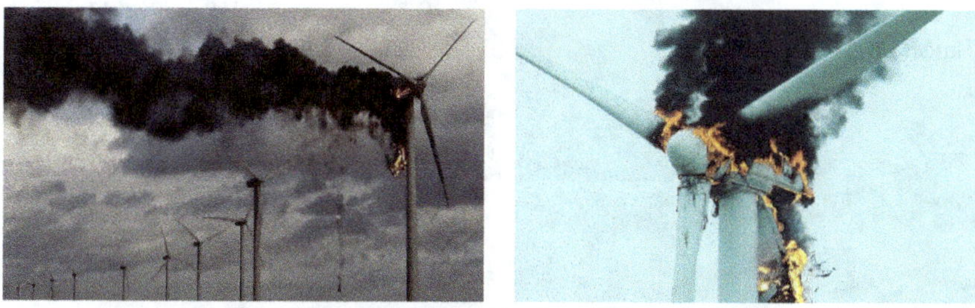

Fig. 1.13 Damage caused by fire (Fire ;Wind Power)

The most dangerous damage is caused by brake failure in high winds. In these conditions, the whole rotor can break loose and strike the tower or nacelle.

There have been known cases where the entire nacelle with the rotor attached has broken loose and hit the tower. Pieces of blades and other sub-assemblies have been thrown up to 500m from the tower. In 2006, Hurricane Samoai hit a wind turbine farm in China. Wind speeds reached 67 m/s and only 8 of the 28 turbines were not damaged. The most serious damage was to the blades of the turbines, which were broken into several pieces. Two 660kW turbines and one 750kW turbine were completely collapsed (H.F.Zhou et al., 2014).

Table 1.2 summarizes the main failure modes of wind turbine blades (Branner, K., 2014).

Table 1.2 Modes of blade failure

Failure mode	Photo	Classification	The result of failure
Interlaminar failure	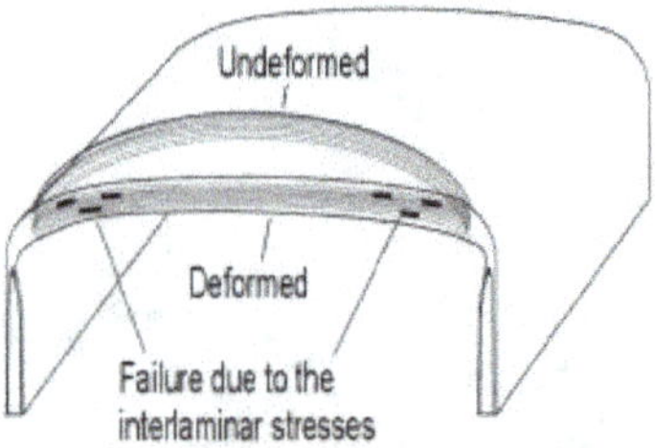	V2-V3	Brazier effect, Bending moment
Delamination - faulty injection	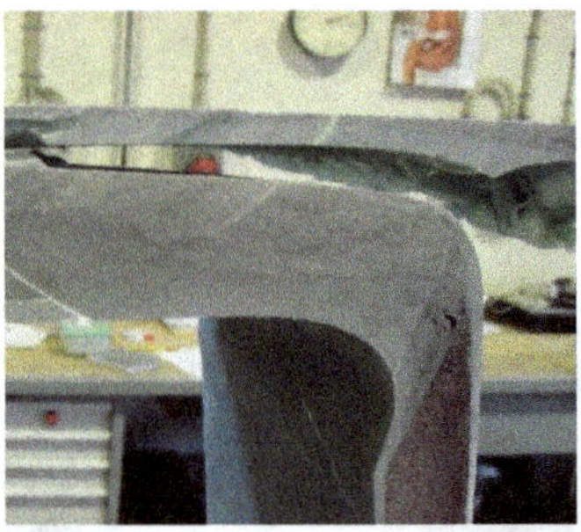	V1	Manufacturing
Peeling / Wear	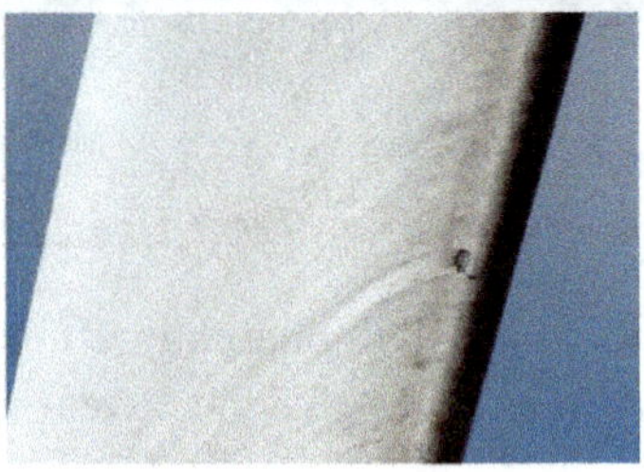	V1	Manufacturing

Flaking of the top coat	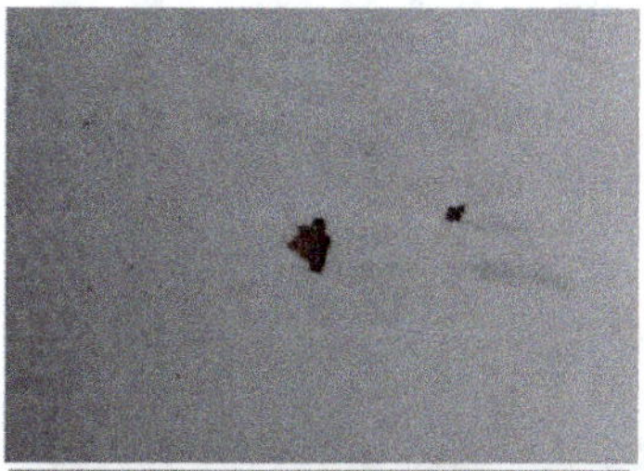	V1	Air bubbles from the manufacturing /poor quality
Missing external parts	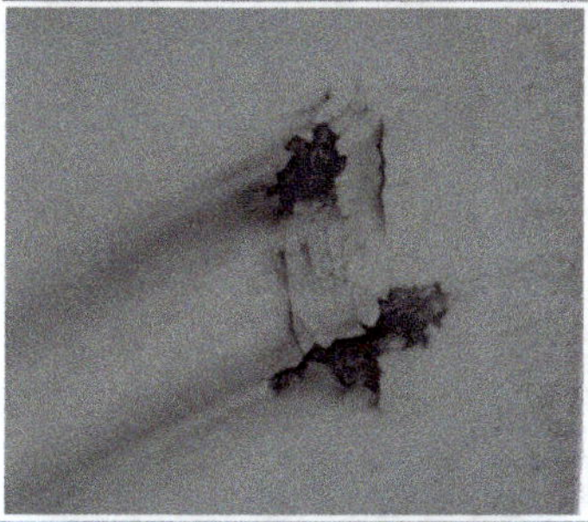	V2-V3	Flaking and external objects impact
Transverse cracks on blade surface	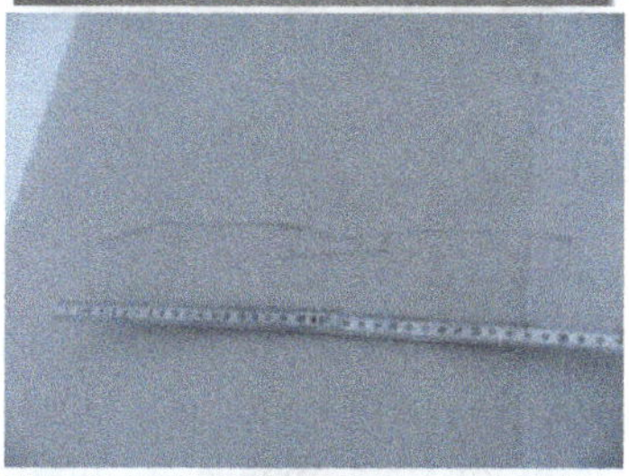	V2-V3	Poor design
Front edge cracks (transverse and longitudinal)			
Fatigue failure in bondlines, longitudinal cracks in the trailing edge		V1-V2	Transversal shear distortion, Deformation of trailing edge panels, Trailing edge buckling

Lightning damage	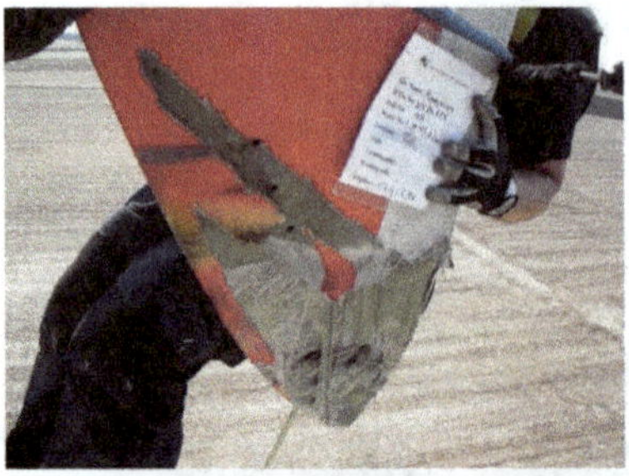	V3	Lightning
Transport damage		V3	

1.4 Composite materials testing

Composite materials with polymer matrix are used widely to manufacture wind turbine blades. These materials have enabled the construction of large diameter rotors. The composite material is a multiphase system obtained artificially or naturally by the association of at least two chemically distinct materials, with a clear separation interface between the two components, with the aim of obtaining superior performance (low weight, increased strength and high durability) compared to the starting components (Bârsănescu P. D. et al., 2004). Long fiber reinforced composites are inhomogeneous, orthotropic materials and the matrix has low hardness. All these characteristics create difficulties when testing composite materials. The number of independent elastic characteristics of the materials is presented in Table 1.5 (Bârsănescu P. D. et al., 2004).

Tabel 1.3 Number of independent elastic characteristics for different categories of MC

	Material			
	Anizotrop	General orthotrop	Orthotrop with transverse isotropy	Isotrop
Number of characteristics	21	9	5	2

Due to the larger number of elastic characteristics, several tests are needed to determine them. There are attempts to determine all elastic characteristics of an orthotropic material by a single test. A disk-shaped specimen has been proposed, subjected to bending with three types of loadings which, combined together, allow the determination of six elastic characteristics of an anisotropic material (Fabrice Pierron, 2023).

The in-plane properties of composite materials have been studied using different tests: T-shaped specimens (Grédiac, M., et al., 1999), short beams subjected to three-point bending (Grédiac, M., et al., 2004; Jiang, L., et al., 2015; He, T., 2018), Iosipescu specimens without notch (Grédiac, M., 1994; Pierron, F., 2000; Chalal, H., et al., 2004; Meraghni, F., 2004; Chalal, H., 2006; Pierron,F., et al., 2007), drilled plates subjected to tensile stress (Gogu, C., et al., 2013; Seon, G., et al., 2019), elliptical-hole plates subjected to tensile stress (Kim, C., 2020).

The shape of the specimens should take into account the shape of the blank from which they were taken. Thick GFRP rings taken from the tubes were used to determine the elastic characteristics in a cylindrical coordinate system (Moulart, R.,2006; Pierron, F., 2000). All orthotropic characteristics of the fabric-reinforced MC that are part of a blade have been identified (Gras, R., et al., 2015).

The DIC (Digital Image Correlation) method has been able to directly identify the mechanical properties of the finished products, which may differ from those of the semi-finished products due to the influence of manufacturing technologies. An additional difficulty occurs due to the relatively large deformations of composite materials (MC). The Hopkinson bar method is not reliable for dynamic testing of the elastic characteristics of MC (Fabrice Pierron, 2023).

Efforts are being made to realize new impact tests using accelerometers and dynamometers, such as the image-based inertial impact (IBII) test (Moulart, R., et al., 2011; Pierron, F., et al., 2014). In this way, it was possible to determine the elastic characteristics both in plane (Fletcher, L., et al., 2019; Parry, S., et al., 2021) and on the thickness of the MC plate (J. Van Blitterswyk, 2018; J. Van Blitterswyk, 2020).

The basic simple tensile tests of MC are standardized: tensile (ASTM D3039, ASTM D 3518), shear (Iosipescu method ASTM D 5379, between-rail shear ASTM D 4255), compression (ASTM D3410), etc. Many composite materials have different mechanical and elastic characteristics in tensile and compression, respectively. Compression testing of MC is difficult because of the possibility of buckling and crushing of the specimen in the areas of contact with the platens or pans of the testing machine. Three types of compression tests are performed: with short specimens, guided in the calibrated portion (type I); with longer specimens, clamped in trays (type II); with sandwich specimens (two flat specimens made of MC, between which a honeycomb-type structure is glued to prevent buckling) (Bârsănescu P. D. and others, 2004). For compression testing of flat MC specimens, the modified ASTM D695 test is commonly used, i.e. Boeing BSS 7260, Figure 1.14.

The compression testing with type I and II specimens has the following disadvantages: the unguided portion of the specimen is small and, for this reason, the installation of a tensile mark is difficult and the use of the DIC method is not possible.

Fig. 1.14 Compression test device according to ASTM D695 modified **(Boeing BSS 7260)**

The materials used in the construction of wind turbine blades should have low density, high resistance to static stress, fatigue and degradation (Ribrant J., 2007; Thomsen OT., 2009). The blade skin that defines the airfoil is typically constructed of GFRP polymer matrix composites (Veers, P., 2003).

The mechanical and elastic characteristics in different directions of the fiberglass-reinforced composite material have been studied and it was found that this material can be successfully used in the manufacture of turbine blades (Goanta V., et al, 2010; Steigmann R., et.al.,). The tensile test results of two types of glass fiber reinforced polymers (one type reinforced with RT500 fabric and the other type reinforced with MAT450 fibers) were presented in the paper (Stanciu, M.D., et.al., 2021). The modulus of elasticity value for GFRP-RT500 in the longitudinal direction was E=21337MPa, twice higher than GFRP-MAT450, which had a value of E=10238 MPa. Tensile, compression and bending tests were performed on GFRP extracted from turbine blades that had been in service for 20 years. It was found that the material properties determined by tensile testing performed on specimens taken from the blade span showed superior properties. The maximum compressive strength was obtained for specimens taken in the transverse direction and the lowest in the blade length direction. The bending properties are significantly affected by fiber orientation, with the best properties obtained for the samples taken from the blade length and the lowest for the samples taken from the transverse direction (Ahmed, M.M.Z., et al., 2021).

The mechanical and elastic characteristics of GFRP, the influence of water/moisture on the internal structure of MC, are currently being investigated by researchers worldwide (Abenojar, J., et al., 2015; Chaichanawong, J., et al., 2016; Choi, HS., et al., 2001; Cordin, M., et al., 2018).

Moisture has been demonstrated to have a significant role in the long-term durability of glass fiber-reinforced polymer matrix composites, influencing the mechanical characteristics of composite materials, see Table 1.3, (Cerbu C, 2007). Results of experimental investigations (Corum, J.M, et al., 2001) have demonstrated degradation of mechanical characteristics due to moisture absorption in unidirectionally or bidirectionally reinforced MC.

Experimentally, water absorption after 4800 h and the degradation of mechanical properties in the transverse direction in the fiber were studied for polymer-matrix MC. It resulted in a reduction of the longitudinal modulus of elasticity E ranging from 55% to 65% (Pomies F., et al., 1995).

Manfredi L.B. et al. studied the effect of water absorption in epoxy matrix composite material and found that water absorption affects both the matrix and the fiber-matrix interface, causing micromechanical damage, which visibly leads to a decrease in the mechanical properties of the material studied (Manfredi L B., et al., 2008). Imielinska K. et al. have studied the behavior of glass fiber reinforced epoxy matrix composite material and obtained results demonstrating that water absorption has a negative effect on the studied material. Thus, the water content affected the microstructural integrity of the material, causing internal defects (Imielinska K., 2004). Exposure of the polymer matrix composite material to aggressive environment caused severe degradations in the fiber-matrix interface, as well as excessive fiber debonding (Ramirez F.A., et al., 2008). It has been shown that the presence of moisture and water particles in the matrix or fiber-matrix interface will lead to reduction in mechanical, chemical and thermophysical properties (Sampath Rao, et al., 2012; Eftekhari, M., et al., 2016; Eslami, S., et al., 20212).

In another study (Lin YC, 2005; Lin YC, 2005; Han, X., et al.,2014), the adsorption-desorption-resorption characteristic of moisture in epoxy system was investigated by hygrothermal aging and molecular dynamics simulation. Due to the effect on hygrothermal aging, Young's modulus and tensile strength were reduced. The effect of moisture absorption on the MC and the results obtained from mechanical tests showed that the negative effects manifested through plasticization and volume size of the matrix, affecting the interfacial reaction between the matrix and the epoxy fiber. Moisture getting into the voids between fiber and matrix causes a loss of adhesion between them (Behzad Ahmed Zai, et al.,2009).

1.5. General conclusions to Chapter 1

1. Developing renewable energy will make a considerable contribution to climate change adaptation. Romania has a high wind potential, but still insufficiently exploited. For these reasons, the development of wind energy will grow rapidly in the future;
2. The blade is the element in a wind turbine that is the least reliable. The breakage of one blade usually causes damage to the other blades. More than 65% of wind turbine accidents between 2015-2020 were due to blade damage. This underlines the importance of correct design and proper testing of blades and the materials from which they are made;
3. Wind turbine blades have a complex structure, the typical cross-section being that of a thin-walled multi-cone multi-cone profile. Several materials are used in the manufacturing of the blades, which are blended together: metal alloys, composites (fiber-reinforced, fabric-reinforced, sandwich), protective materials (primer, UV protection gels, paint, etc.). Both the component materials and the entire structure of the blade must be tested to establish the limiting characteristics.
4. Composite materials used in turbine blades (GFRP and CFRP) absorb moisture from the environment and this leads to a decrease in mechanical and elastic characteristics. For this reason, it is necessary to test these materials both dry and after moisture absorption. The determination of the elastic and mechanical properties of composite materials is done by standardized tests.

CHAPTER 2. EXPERIMENTAL RESEARCH ON THE CHARACTERISTICS OF THE MATERIAL FROM WHICH

2.1 The purpose of experimental research on GFRP material

Glass fiber reinforced composite (GFRP) is used as the predominant material in the manufacture of wind turbine blades, as presented in Chapter 1. The aim of the experimental research is to provide information on the mechanical and elastic characteristics of the material used in the manufacture of wind turbine blades. Accurate knowledge of the material properties is very important, because during operation the element most prone to damage is the blade.

2.2 Development of GFRP composite material

In order to determine the material characteristics, two plates of RT 500 fiberglass reinforced composite were manufactured, the fiber impregnation being carried out with EPIKOTE MGS LR 385 epoxy resin matrix by the lay-up method. The resin characteristics are shown in Table 1 (EPICOTETM Resin MGS® LR 385).

Table 2.1 Characteristics of epoxy resin type EPIKOTE MGS LR 385

Properties	Units	Value
Density	[g/cm^3]	1,20
Viscosity	[mPa×s]	700-1050
Flexural strength	[MPa]	120-130
Modulus of elasticity (Young)	[MPa]	3,3-3,6
Tensile strength	[MPa]	75-85
Compressive strength	[MPa]	120-140
Elongation of break	[%]	6-8
Impact strength	[KJ/m^2]	45-60
Water absorption at 23°C in 24h	[%]	0,01

The first composite plate is reinforced with 10 layers of bidirectional fabric, as in Figure 2.1a, with fiber orientation at [0°/90°]. Specimens were cut from the plate using a diamond disk cutter, limiting the heating in the cutting area and thus avoiding edge effects (delaminations and fiber pull-outs). After the cutting process, the samples were finished and a total of fifteen specimens taken in three different directions were obtained: 5 transverse (TR) in the direction of the weft, 5 longitudinal (LG) in the direction of the warp and 5 diagonally at 45° (45), Figure 2.1b.

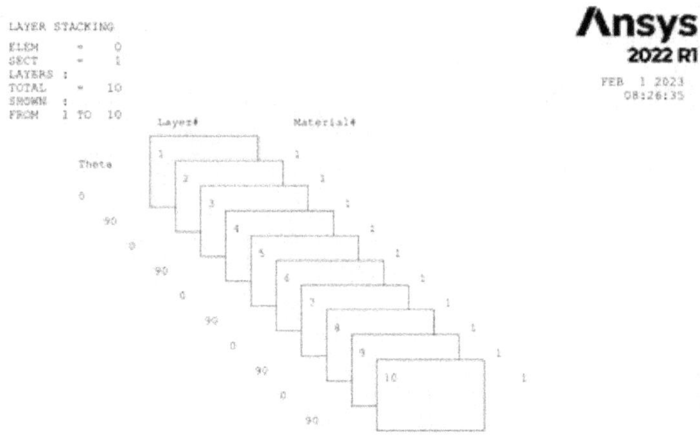

a)

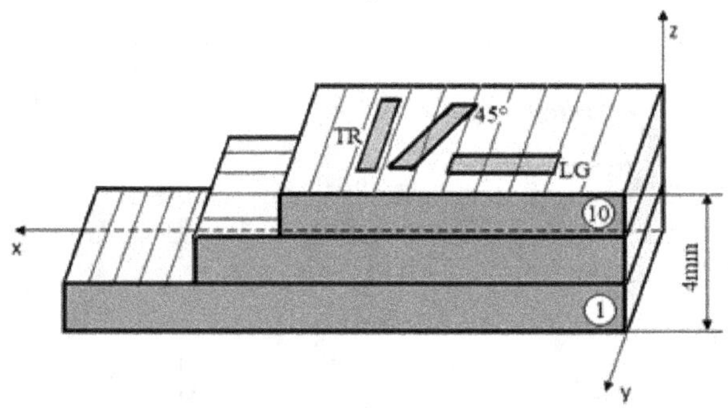

b)

Fig. 2.1 a) Storage sequence of layers [0°/90°] and directions for sampling from the GFRP plate b)

The second plate has the same number of layers as the first, but the fiber orientation is at [-45°/0°/+45°/90°], Figure 2.2a, and from which ten specimens were cut in two directions (5 transverse (TR), 5 longitudinal (LG)), Figure 2.2b.

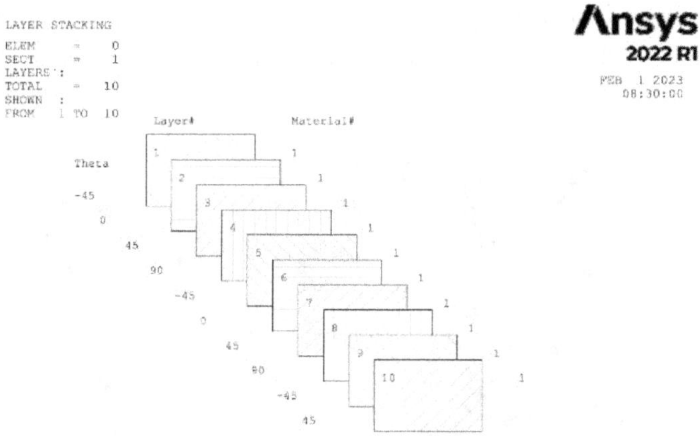

a)

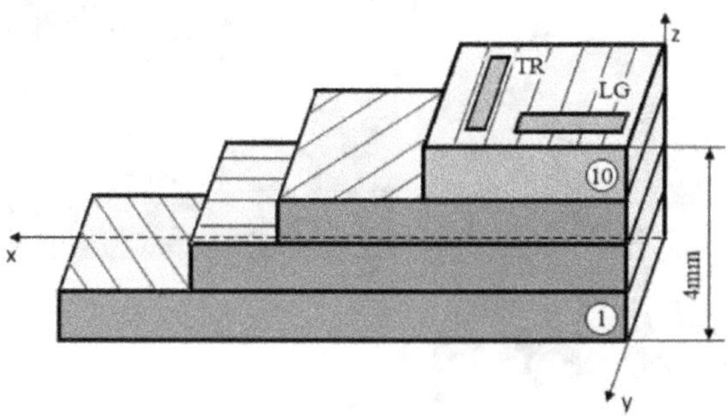

b)

Fig. 2.2 a) Storage sequence of layers [-45°/0°/+45°/90°] and directions for sampling from the GFRP plate b)

2.3 Tensile Tests of GFRP

Tensile testing of composite specimens, which were clamped in the universal testing machine (INSTRON 8801) in order to determine the material characteristics, was carried out in accordance with ASTM D 3039.

The loading speed was 0.5mm/min for each specimen. Due to the low speed of displacement during the test, it was possible to observe the deformations and damages occurring, thus obtaining a good accuracy of the characteristic curve tracing. The pressure in the hydraulic clamping system of the machine's troughs was 70 bar, adapted to the material on the sample surface. The gripping cups must achieve the necessary clamping for the test but must not crush the material in the contact area. The strains in the stress direction were measured using an extensometer of the type "Dynamic Extensometer -2670-112" with an initial length between the marks of 25 mm, Figure 2.3.

Fig. 2.3 Tensile test and details with INSTRON -2670-112 extensometer positioning

The specimens were made according to ASTM D 3039, Figure 2.4.

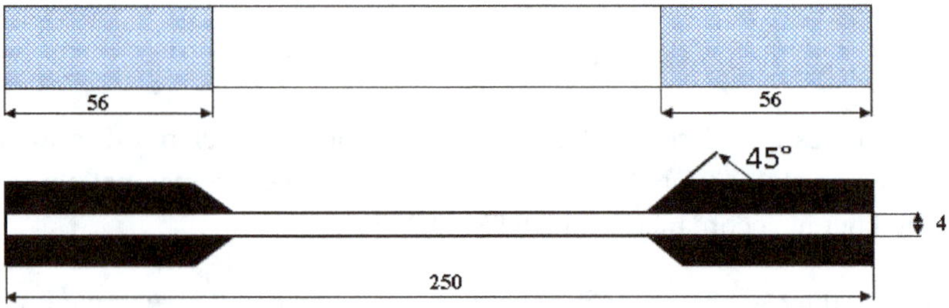

Fig. 2.4 GFRP specimen dimensions according to ASTM D3039

On the portion where the specimen was fixed in the machine's gripping trays, reinforcing plugs made of the same material as the specimens were made of were glued to avoid damage and breakage in the gripping trays. The bead weaving at 45° was done in order to reduce the stresses in the area of the end-pressing of the beads on the specimen. The gluing of the beads was performed according to the standard specified above. Specimens were cut from the plate with fiber orientation at [-45°/0°/+45°/90°]: four each in longitudinal and transverse direction. From the [0°/90°] fiber orientation plate, four specimens each were obtained in the longitudinal direction (LG) in the warp direction, transverse direction (TR) in the weft direction and diagonal direction at 45°, Figure 2.5.

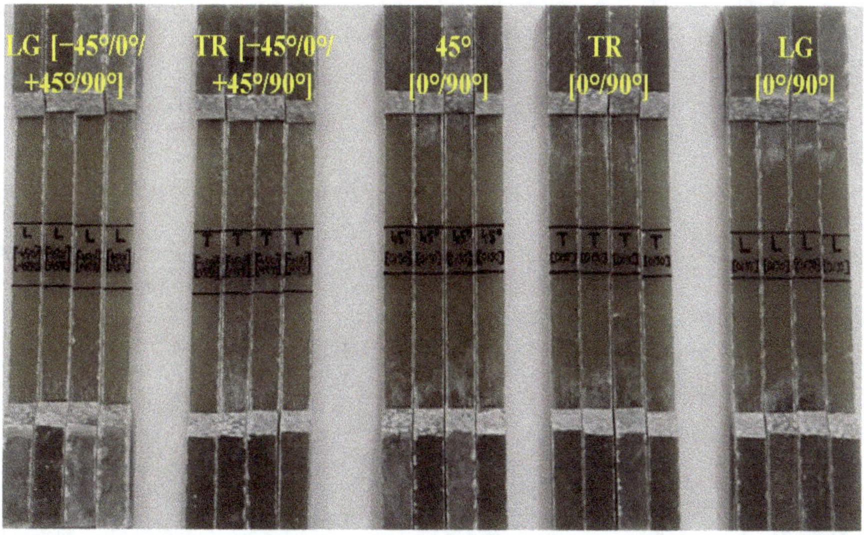

Fig. 2.5 Specimens used for tensile testing

On the basis of the results obtained after failure of the specimens, specific stress-strain characteristic curves were plotted for all the specimens taken from the two plates. Given the large number of results obtained, a statistical analysis of the data was also performed. The formulas used were the same for all five cases of specimen sets. In addition, for each set of specimens, the failure modes were also analyzed, Figure 2.6.

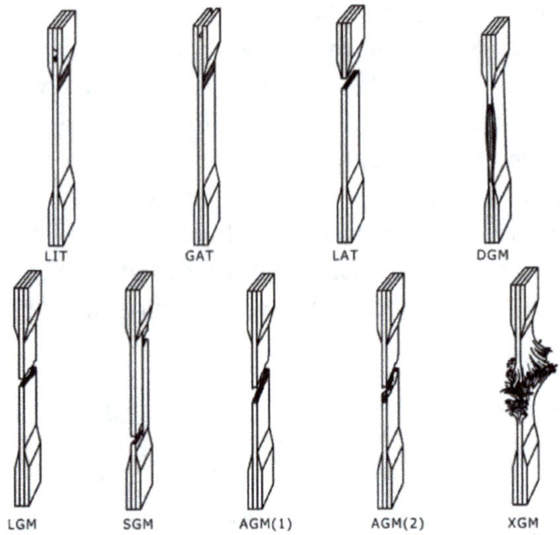

Fig. 2.6 Tensile test failure codes/typical failure modes according to ASTM D 3039

Figure 2.7 shows the overlaid characteristic curves for the 4 specimens cut longitudinally (in the warp direction) from the plate with fiber orientation at [0°/90°].

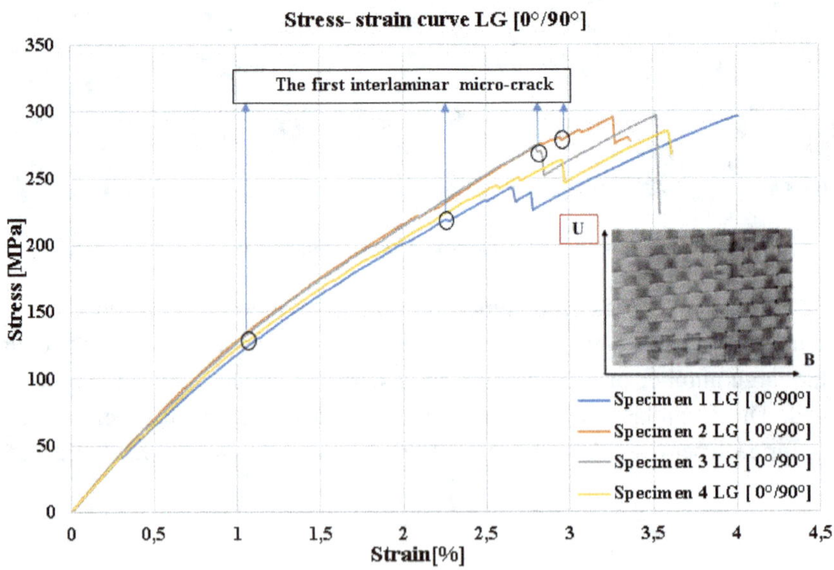

Fig. 2.7 Stress-strain diagram for longitudinally cut specimens at [0°/90°]

A comparison of the 4 curves shows that the maximum tensile stress occurs at specimen 3, with ultimate tensile stress of σ_{UTS}= 296,37 MPa. By analyzing the characteristic stress-strain curves, it can be seen that between the elastic limit and the fracture zone of the material there is a first portion that indicates the interlaminar behavior and the response of the outer laminate during stress (most of the tensile stress is taken up by the fibers in the direction of the applied force). The failure initiation is indicated by the appearance of interlaminar microcracks in the material, which cause jumps in the characteristic diagram. On each of the four curves the appearance of a first interlaminar microcrack was identified, after which the tensile stress was redistributed to the rest of the composite As the force increased, more and more fibers failed and the load was transferred to the fibers that did not fail. Due to delaminations occurring between the reinforcing fibers and the matrix, they no longer work together and the damage propagates into the material. This phenomenon leads to the rapid spread of the failure and the breakdown of the composite. In specimen 1, the first microcrack was initiated at a stress of approximately σ_r= 220 MPa, in specimen 2 at σ_r= 280 MPa, in specimen 3 at σ_r= 275 MPa and in specimen 4 the first crack was observed at a stress of approximately σ_r= 120 MPa. By carefully analyzing the four slopes of the stress-strain curves, it was found that little strain energy storage (area under the curve) is exhibited.

The following is the standard deviation for the four-tensile breaking tensile stresses σ_r (longitudinally cut from the plate with fiber orientation at [0°/90°], Table 2.2.

Table 2.2 Standard deviation for ultimate tensile stress σ_{UTS}, (longitudinal oriented at [0°/90°]).

Sample no.	σ_{UTS} [MPa]	$\bar{\sigma}_{UTS}$ [MPa]	Deviations from the mean $\sigma_r - \bar{\sigma}_r$ [MPa]	$(\sigma_r - \bar{\sigma}_r)^2$ [MPa]	Standard deviation S [MPa]	Coefficient of variation CV [%]
1	296,03	293,2	2,83	8,00	5,16	1,75
2	294,93		1,73	2,99		
3	**296,37**		3,17	10,04		
4	285,52		-7,68	58,98		
Σ=	1172,8		-	80,01		

The arithmetic mean of these four data points is the average ultimate tensile stress:

$$\bar{\sigma}_r = \frac{1172,8}{4} = 293,2MPa$$

The sample standard deviation is:

$$S = \sqrt{\frac{\sum(\sigma_r - \bar{\sigma}_r)^2}{n-1}}; S = \sqrt{\frac{80{,}01}{3}} = 5{,}16MPa$$

The coefficient of variation CV is defined as the ratio of the standard deviation S to the mean:

$$CV = \frac{S}{\bar{\sigma}_r}; CV = \frac{5{,}16}{293{,}2}100 = 1{,}75\%$$

At the same time, the failure of the specimens was analyzed, which is presented in Table 2.3.

Table 2.3 Failure mode of specimens taken longitudinally from the plate [0°/90°], according to ASTM D 3039.

Sample no.	Failure mode according to ASTM D3039 (Fig. 2.6)	First character	Second character	Third character
1	DGB	Delamination	Gage	Botton
2	DGB	Delamination	Gage	Botton
3	DGM	Delamination	Gage	Middle
4	DWB	Delamination	<1W from grip/tab	Botton

Figure 2.8 showed the failure mode for each specimen. In all four cases, the degradation mechanism occurred as interlaminar cracks. The area of delamination breakage for specimens 1, 2 and 3 occurred in the calibrated area and for specimen 4 near the tab. The breakage location for specimens 1, 2 and 4 was at the bottom, and for specimen 3 in the middle of the calibrated area.

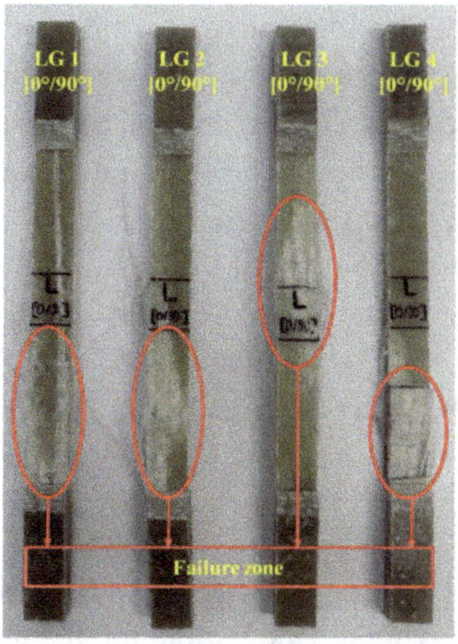

Fig. 2.8 Failure specimens taken longitudinally, with fiber orientation at [0°/90°]

Figure 2.9 shows the overlaid characteristic curves. The differences in tensile behavior for the four specimens cut transversely from the plate with the fiber orientation at [0°/90°] are shown.

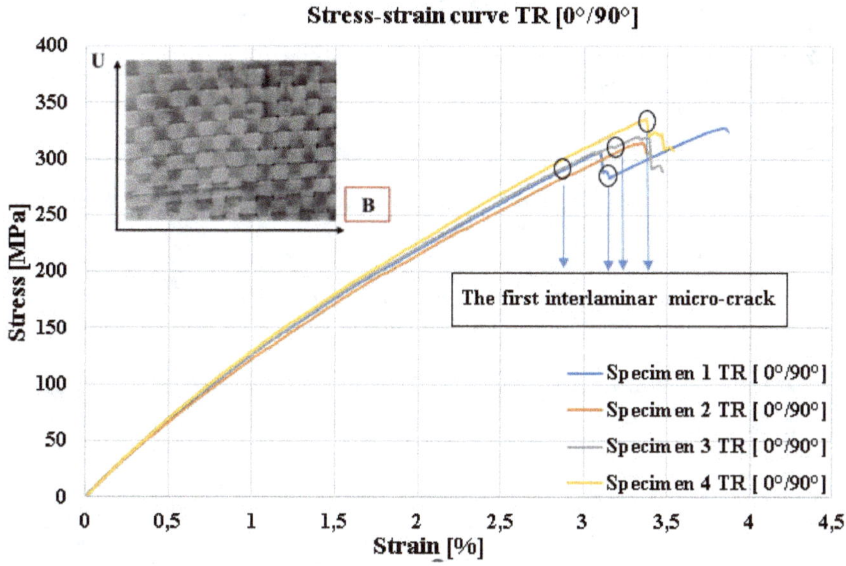

Fig. 2.9 Stress-strain diagram for transversal cut specimens at [0°/90°]

A detailed analysis of the four curves reveals that the ultimate tensile stress σ_{UTS}= 334.87 MPa occurs at specimen 4. Analysis of the characteristic curves shows that there is a first step between the elastic zone and the fracture zone of each specimen, which indicates interlaminar failure. The first interlaminar micro-cracks initiated during the test were identified on the four curves. After the first crack appeared, the normal stress was distributed in the rest of the material. As the force increases, more fibers break and the load is transferred to the fibers that have not failed and thus take up the entire load. Thus, the bond between the reinforcing fibers and the matrix usually fails first. In the case of the four specimens, the first micro-crack was initiated at stresses between σ_r= 290 MPa and 330 MPa. Also in the case of the cross-cut specimens, the slopes of the specific stress-strain curves show little strain energy storage (area under the curve).

The following is the standard deviation for the four-tensile breaking tensile stresses σ_r (transversal cut from the plate with fiber orientation at [0°/90°], Table 2.4.

Table 2.4 Standard deviation for ultimate tensile stress σ_{UTS}, (transversal oriented at [0°/90°]).

Sample no.	σ_{UTS} [MPa]	$\bar{\sigma}_{UTS}$ [MPa]	Deviations from the mean $\sigma_r - \bar{\sigma}_r$ [MPa]	$(\sigma_r - \bar{\sigma}_r)^2$ [MPa]	Standard deviation S [MPa]	Coefficien t of variation CV [%]
1	327,47		3,28	10,75		
2	314,27		-9,92	98,04		
3	320,17	324,19	-4,02	16,16	8,92	2,75
4	**334,87**		10,68	114,06		
Σ=	1296,7		-	239,01		

At the same time, the failure of the specimens was analyzed, which is presented in Table 2.5.

Table 2.5 Failure mode of specimens taken transversal from the plate [0°/90°], according to ASTM D 3039.

Sample no.	Failure mode according to ASTM D3039 (Fig. 2.6)	First character	Second character	Third character
1	DGM	Delamination	Gage	Middle
2	DGM	Delamination	Gage	Middle
3	DGM	Delamination	Gage	Middle
4	DAT	Delamination	At grip/tab	Top

Figure 2.10 shows the failure mode of each specimen. In all four cases, the failure mechanism was the initiation of interlaminar cracks. Delamination failure localization for specimens 1, 2 and 3 occurred in the calibrated area and for specimen 4 near the tab. The fracture location for specimens 1, 2 and 3 was noted in the middle and for specimen 4 at the top.

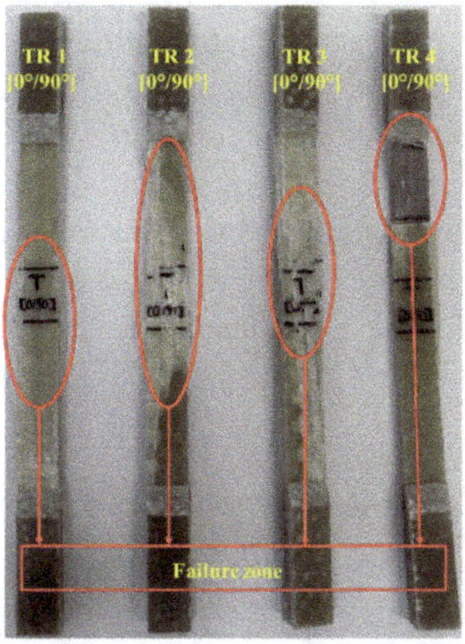

Fig. 2.10 Failure specimens taken transversal, with fiber orientation at [0°/90°]

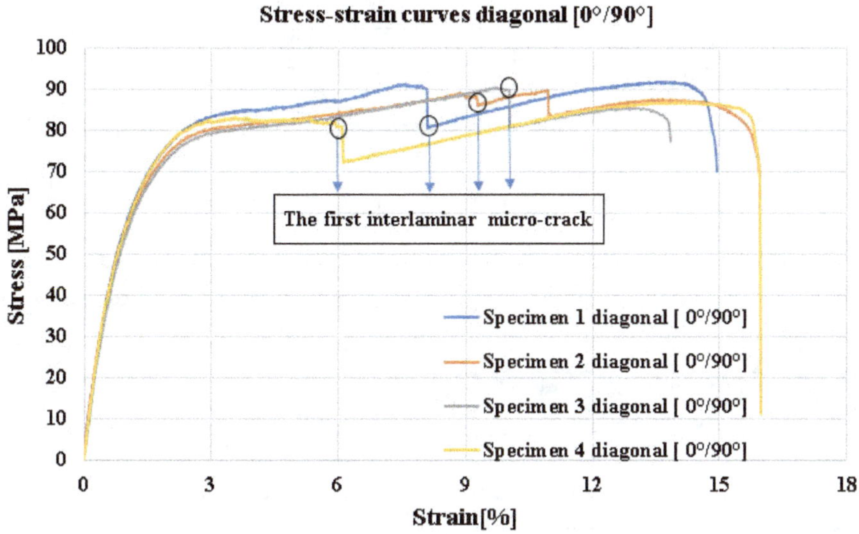

Fig. 2.11 Stress-strain diagram for diagonal cut specimens at [0°/90°]

The overlaid stress-strain curves in Figure 2.11 show the differences in the tensile fracture toughness behavior for the 4 specimens cut at 45° from the plate with fiber orientation at [0°/90°]. In the case of specimens cut at 45°, the laminates are no longer stressed in the direction of load application, and the strength characteristics of the resin become important in the structural analysis of the layered composite material. With respect to the specimens loaded in the longitudinal and transverse direction, a deviation from linearity of the specific strain stress-strain curves and much lower values of the fracture toughness are observed. Analyzing the 4 curves, it can be observed that the average the ultimate tensile stress is σ_{UTS}=89,70 MPa. Following carefully the shape of the specific stress-strain curves, it was found that they show a high deformation energy storage, this aspect shows an elasto-plastic behavior. This behavior is due to the fact that the stress occurs in a direction along which there are no reinforcing fibers. The following is the standard deviation for the four-tensile breaking tensile stresses σ_r (diagonal cut from the plate with fiber orientation at [0°/90°], Table 2.6.

Table2. 6 Standard deviation for ultimate tensile stress σ_{UTS}, (diagonal oriented at [0°/90°]).

Sample no.	σ_{UTS} [MPa]	$\bar{\sigma}_{UTS}$ [MPa]	Deviations from the mean $\sigma_r - \bar{\sigma}_r$ [MPa]	$(\sigma_r - \bar{\sigma}_r)^2$ [MPa]	Standard deviation S [MPa]	Coefficient of variation CV [%]
1	**91,82**	89,70	2,12	4,49	2,1	2,35
2	89,72		0,02	0,004		
3	90,46		0,76	0,57		
4	86,82		-2,88	8,29		
Σ=	358,82		-	13,35		

At the same time, the failure of the specimens was analyzed, which is presented in Table 2.7.

Table 2.7 Failure mode of specimens taken diagonal from the plate [0°/90°], according to ASTM D 3039.

Sample no.	Failure mode according to ASTM D3039 (Fig. 2.6)	First character	Second character	Third character
1	DWT	Delamination	<1Wf from grip/tab	Top
2	DGT	Delamination	Gage	Top
3	DGB	Delamination	Gage	Bottom
4	DGM	Delamination	Gage	Middle

Figure 2.12 showed the failure mode for each specimen. In all 4 cases, the degradation mechanism occurred in the form of delamination cracks, in which the breakage occurred mainly due to the separation of the layers from each other. This type of breakage generally occurs due to active interlaminar stresses in the matrix layer. The delamination fracture zone for specimens 2, 3 and 4 was in the gauge zone and for specimen 1 near the tab. The breakage location for specimens 1 and 2 was noted at the top, for specimen 3 at the bottom, and for the last specimen it was localized in the middle.

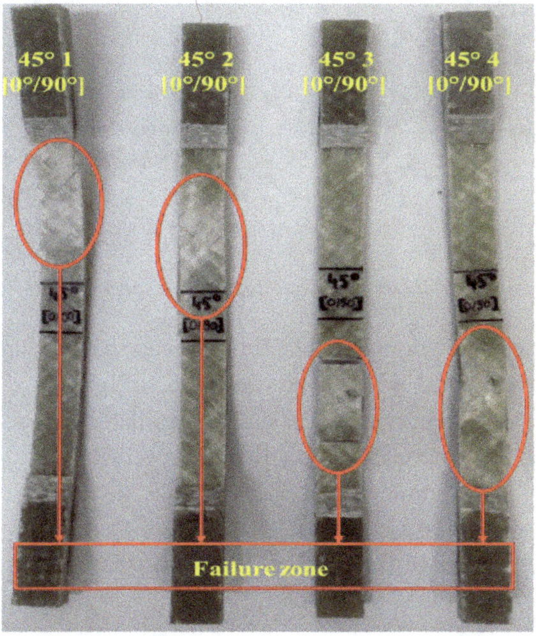

Fig. 2.12 Failure specimens taken diagonal, with fiber orientation at [0°/90°]

Figure 2.13 showed the overlaid specific stress-strain curves and the differences in tensile strength behavior for the 4 longitudinally cut specimens from the plate with fiber orientation at [-45°/0°/+45°/90°].

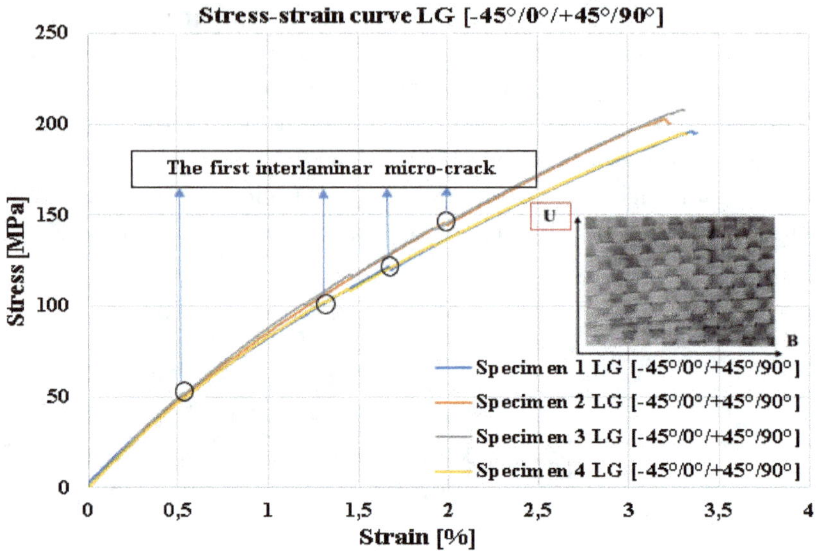

Fig. 2.13 Stress-strain diagram for longitudinal cut specimens at [-45°/0°/+45°/90°]

By comparing the 4 curves, it can be observed that the highest tensile stress at failure occurs at specimen 3 with a maximum value of σ_r= 207.28 MPa. By following each specific stress-strain curve, it can be observed that between the elastic limit and the failure zone of the material there is a first micro-crack which initiated in the range of 55- 140 MPa. After initiation of the first micro-crack, the tensile stress was distributed to the rest of the composite. Moreover, in these diagrams it was found that a small strain energy storage is exhibited, this aspect being specific to materials with visco-elastic behavior.

The following is the standard deviation for the four-tensile breaking tensile stresses σ_r (longitudinal cut from the plate with fiber orientation at [-45°/0°/+45°/90°]), Table 2.8.

Table 2.8 Standard deviation for ultimate tensile stress σ_{UTS}, (longitudinal oriented at [-45°/0°/+45°/90°])

Sample no.	σ_{UTS} [MPa]	$\bar{\sigma}_{UTS}$ [MPa]	Deviations from the mean $\sigma_r - \bar{\sigma}_r$ [MPa]	$(\sigma_r - \bar{\sigma}_r)^2$ [MPa]	Standard deviation S [MPa]	Coefficient of variation CV [%]
1	196,17	200.1	-3,93	15,44	5,95	2,97
2	202,62		2,52	6,35		
3	**207,28**		7,18	51,55		
4	194,35		-5,75	33,06		
Σ=	800,42		-	106,4		

At the same time, the failure of the specimens was analyzed, which is presented in Table 2.9

Table 2.9 Failure mode of specimens taken longitudinal from the plate [-45°/0°/+45°/90°] according to ASTM D 3039.

Sample no.	Failure mode according to ASTM D3039 (Fig. 2.6)	First character	Second character	Third character
1	DGB	Delamination	Gage	Bottom
2	DWB	Delamination	<1Wf from grip/tab	Bottom
3	DGT	Delamination	Gage	Top
4	DWB	Delamination	<1Wf from grip/tab	Bottom

Figure 2.14 shows the failure modes for each sample.

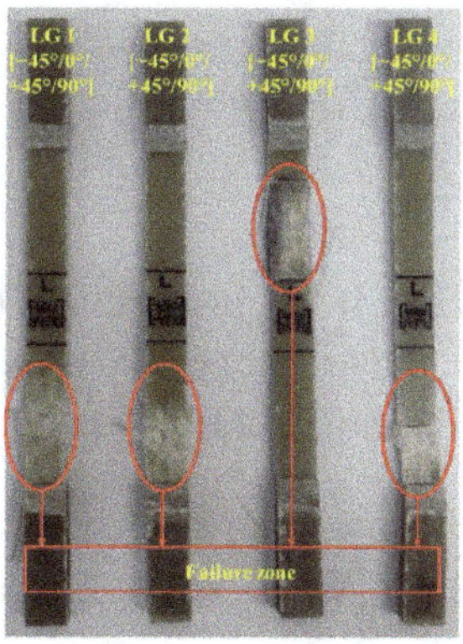

Fig. 2.14 Failure specimens taken longitudinal with fiber orientation at [-45°/0°/+45°/90°]

The degradation mechanism in the four specimens occurred in the form of delamination cracks. The area of delamination breakage for specimens 1 and 3 were in the gauge area, for specimens 2 and 4 near the fillet. The fracture site for specimens 1, 2 and 4 was observed in the bottom and for specimen 3 in the top.

Figure 2.15 shows the overlaid specific stress-strain curves and the differences in tensile fracture toughness behavior for the 4 specimens cross-cut from the plate with fiber orientation at [-45°/0°/+45°/90°].

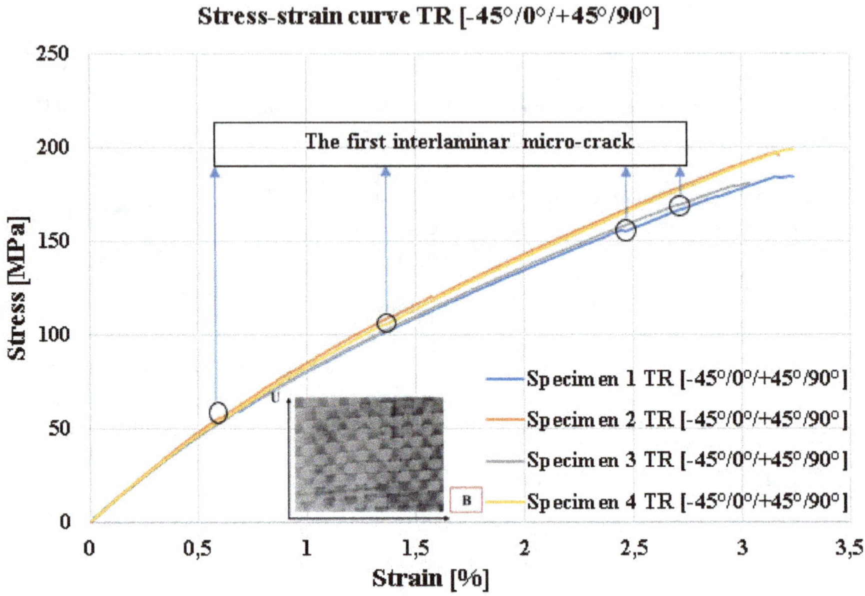

Fig. 2.15 Stress-strain diagram for transversal cut specimens at [-45°/0°/+45°/90°]

By comparing the 4 curves, it can be observed that the biggest tensile stress at failure occurs at specimen 4, having a maximum value of σ_r= 198.28 MPa. By analyzing the stress-strain curves, high similarities can be observed between the elastic limit and the fracture zone of the 4 specimens as in the case of longitudinally cut specimens from the same plate. We find a visco-elastic behavior of the material due to the small deformation energy storage under the characteristic curves.

The following is the standard deviation for the four-tensile breaking tensile stresses σ_r (transversal cut from the plate with fiber orientation at [-45°/0°/+45°/90°]), Table 2.10.

Table 2.10 Standard deviation for ultimate tensile stress σ_{UTS}, (transversal oriented at [-45°/0°/+45°/90°])

Sample no.	σ_{UTS} [MPa]	$\bar{\sigma}_{UTS}$ [MPa]	Deviations from the mean $\sigma_r - \bar{\sigma}_r$ [MPa]	$(\sigma_r - \bar{\sigma}_r)^2$ [MPa]	Standard deviation S [MPa]	Coefficient of variation CV [%]
1	184,17	188,44	-4,27	18,23	9,12	4,83
2	196,87		8,43	71,06		
3	180,45		-7,99	63,84		
4	**198,28**		9,84	96,82		
Σ=	753,77		-	249,95		

Table 2.11 shows the failure mode of specimens transverse cut from the plate with fiber orientation at [-45°/0°/+45°/90°].

Table 2.11 Failure mode of specimens taken transversal from the plate [-45°/0°/+45°/90°] according to ASTM D 3039.

Sample no.	Failure mode according to ASTM D3039 (Fig. 2.6)	First character	Second character	Third character
1	DGB	Delamination	Gage	Bottom
2	DAB	Delamination	At grip/tab	Bottom
3	DAB	Delamination	At grip/tab	Bottom
4	DWB	Delamination	<1Wf from grip/tab	Bottom

Figure 2.16 showed the failure mode for each specimen.

In all 4 cases, the degradation mechanism occurred in the form of delamination cracks, in which the failure occurred mainly due to the separation of the layers from each other. The delamination breakage zone for specimen 1 occurred in the gauged zone, for specimens 2, 3 and 4 near the table. The breakage location for specimens 1, 2, 3 and 4 was at the bottom.

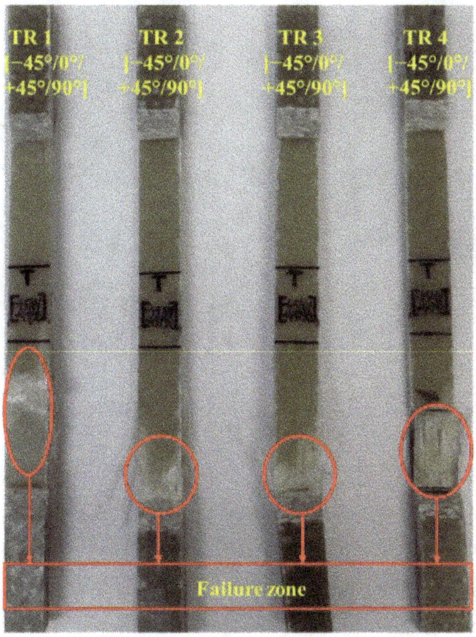

Fig. 2.16 Failure specimens taken transversal with fiber orientation at [-45°/0°/+45°/90°]

2.4 Determination of elastic characteristics of *GFRP* composite material

Unidirectionally reinforced composite materials belong to the category of orthotropic materials, which have three planes of symmetry in their composition (Hadăr, A, 1997), (Alămoreanu, E., et al., 1993). A structure made of layered and reinforced MCs with continuous fibers requires the study of the elastic characteristics of a laminate:

- E_{11}- longitudinal modulus of elasticity (Young);
- E_{12}- transverse modulus of elasticity;
- G_{12}- shear elastic modulus (Coulomb);
- ν_{12}- Poisson coefficient.

The elastic and mechanical characteristics of the material were determined by the resistive electrical tensiometry method (Bârsănescu P.D. et al., 2004). Figure 2.17 shows the specimens cut from the two plates and the strain gauges location.

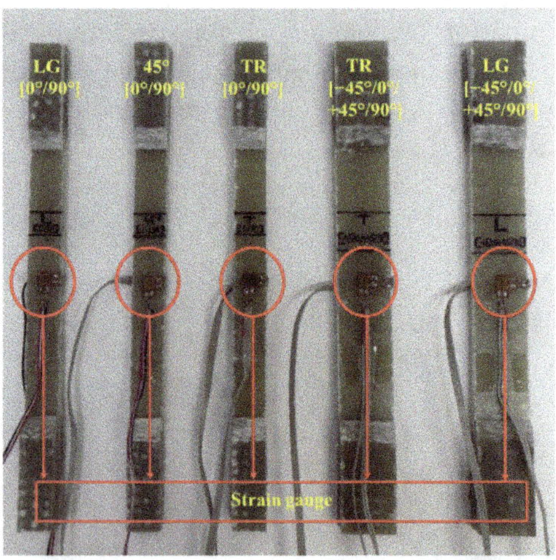

Fig. 2. 17 Position of T-type strain gauges

For the specimens taken from the [0°/90°] plate, three specimens were cut in the three known directions, following the ASTM D3039 standard, presented in subchapter 2.2. For specimens with fiber orientation at [-45°/0°/+45°/90°], the ASTM D3518 standard, Figure 2.18, was followed, from which two specimens were cut: longitudinal and transverse direction.

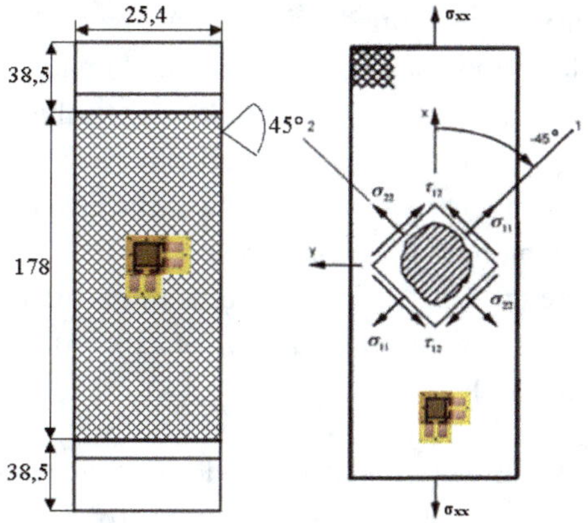

Fig. 2.18 Position of T-type strain gauges and the stress state of the specimen

In order to obtain the elastic and mechanical characteristics, a T-type strain gauge (with two grids at 90°) of the type CEA-06-125WT-120, manufactured by Micro-Measurements, was bonded to each specimen. The characteristics of the strain gauge used in the tensile tests are as follows: resistance R=120Ω±0,35% and gauge factor: k_G=2,025±0,5%. The strain gauge was installed according to the manufacturer's recommendations [http:// www. vishaypg.com/micro-measurements]. The mounting configuration was quarter-bridge, Figure 2.19.

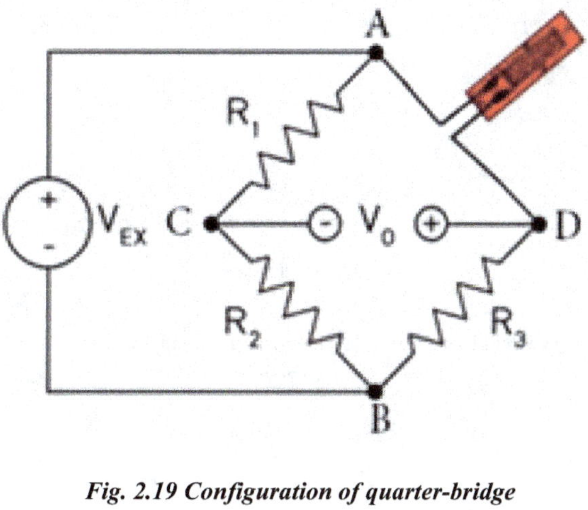

Fig. 2.19 Configuration of quarter-bridge

Data acquisition was performed using the Vishay P3 tensiometer bridge. The results obtained after load application, in the elastic domain, were read on the bridge display and the experimental results were recorded for further processing. Under these conditions, the strain gauge in the longitudinal direction will measure the specific longitudinal strain, ε_L and the strain gauge in the transverse direction will measure ε_T. The results obtained are plotted in normal stress (σ) / specific longitudinal strain (ε_L) coordinates to determine the longitudinal modulus of elasticity, E, (Young's modulus) for the plate specimens with orientation at [0°/90°], Figure 2.20 and [-45°/0°/+45°/90°], Figure 2.21.

Young's modulus was determined as the slope of the straight-line approximation of the graph plotted in the coordinates normal stress (σ) / specific longitudinal strain (ε_L), through the points determined from the Instron 8801 machine computer records of force values and specific longitudinal strains.

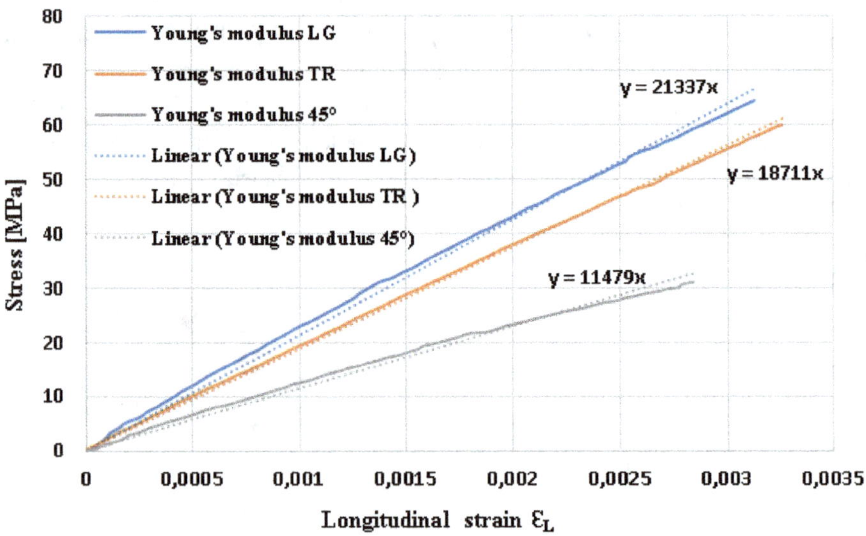

Fig. 2.20 Stress variation with longitudinal strain-approximation lines for longitudinal elasticity modulus at [0°/90°]

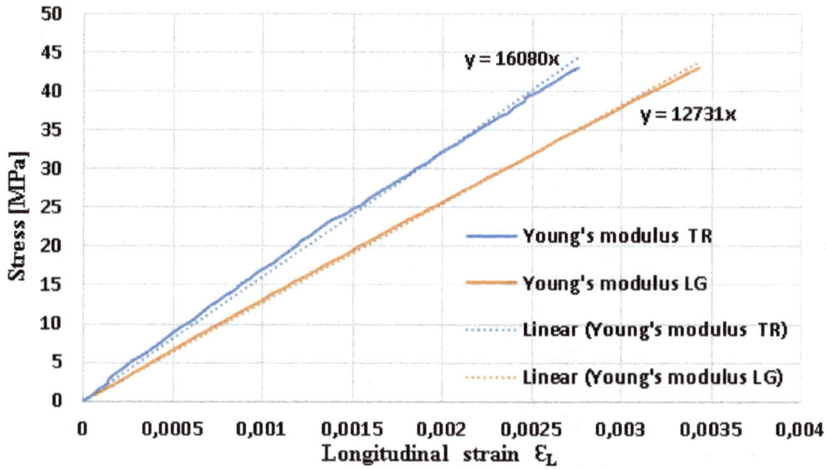

Fig. 2.21 Stress variation with longitudinal strain-approximation lines for longitudinal elasticity modulus at [-45°/0°/+45°/90°]

Figures 2.22 and 2.23 show the transverse specific strain (ε_T) / longitudinal specific strain (ε_L) coordinate plots for the determination of the Poisson's ratio ν, on the plate specimens with orientation at [0°/90°] and [-45°/0°/+45°/90°]. The slope of the approximation line represents the transverse Poisson's contraction coefficient.

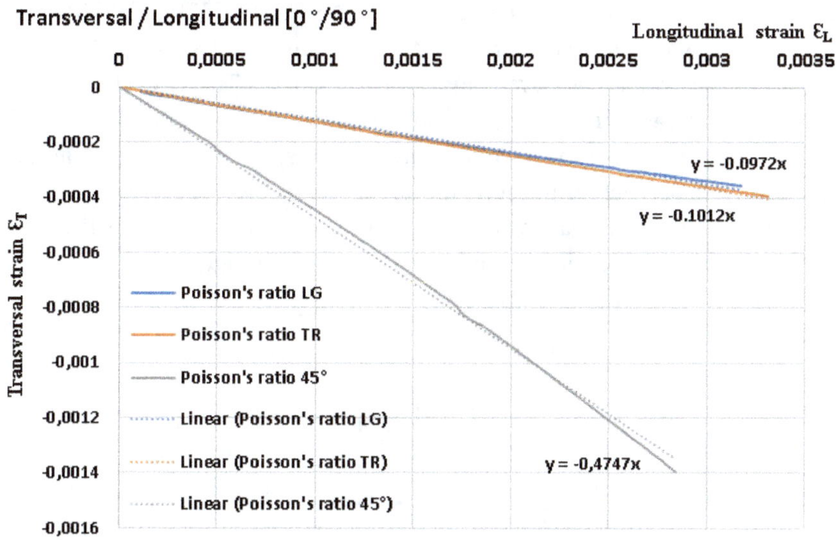

Fig. 2.22 Longitudinal strain-transversal strain curve at [0°/90°]

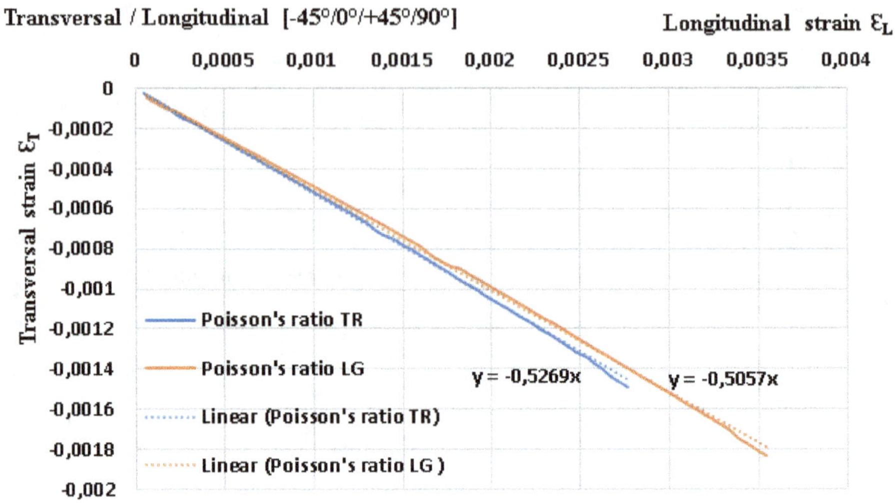

Fig. 2.23 Longitudinal strain-transversal strain curve at [-45°/0°/+45°/90°]

2.5 Shear test

Shear tests were carried out on reinforced specimens at [0°/90°] and [-45°/0°/+45°/90°] orientation according to ASTM D5379/D5379M. For the determination of the elastic characteristics, strain gauges of type C032A, Figure 2.24, manufactured by Micro-Measurements, were mounted. The two strain gauges are arranged on either side of the shear section and are oriented at 90° to each other (T-type). This configuration allows the measurement of specific strains (ε_1 and ε_2) in the vicinity of the shear section in the main directions of the strain/stress state, and the arrangement of the strain gauge in Wheatstone half-bridge structure leads to direct acquisition of the specific shear strain at the Vishay P3 bridge.

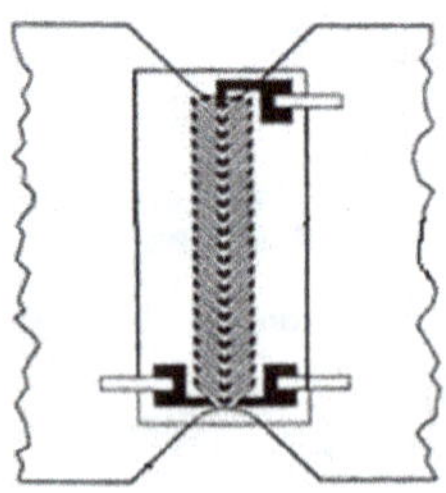

Fig. 2.24 Strain gauges rosette of type C032A

The characteristics of strain gauge are: Resistance R=500Ω, Gauge factor k_G=2,03±1,0% and Transverse sensitivity factor k_T= (+1,2±0,2) %.

Specific shear strain calculus is corrected in function of transverse sensibility of the gauge transducers:

$$\gamma_{1,2} = (\varepsilon_1 - \varepsilon_2) = \frac{1-\nu_0 K_T}{1+K_T}(\varepsilon_{cit.1} - \varepsilon_{cit.2}) = \frac{1-\nu_0 K_T}{1+K_T} \cdot \gamma_{12.cit}$$

Where ε_1, $\varepsilon 2$ şi $\varepsilon_{cit.1}$, $\varepsilon_{cit.2}$ are corrected extensional strains, and respectively effective/read extensional strains. The correction factor is given by the producer of the transducer:

$$\frac{1-\nu_0 K_T}{1+K_T} = \frac{1-0,285 \cdot 0,0019}{1+0,0019} = 0,997563$$

The tests were carried out on a shear device designed at UTIaşi, inspired by the device developed at the University of Idaho (Conant, R.R,1995). The tests determined the shear strengths $\tau_{rf.90}$, $\tau_{rf.45}$ and the transverse modulus of elasticity G_{90}, G_{45}, (Morăraş C., et.al., 2020).

Figures 2.25 and 2.26 show the shear device, installed on the INSTRON 8801 mechanical testing machine, with the two types of specimens.

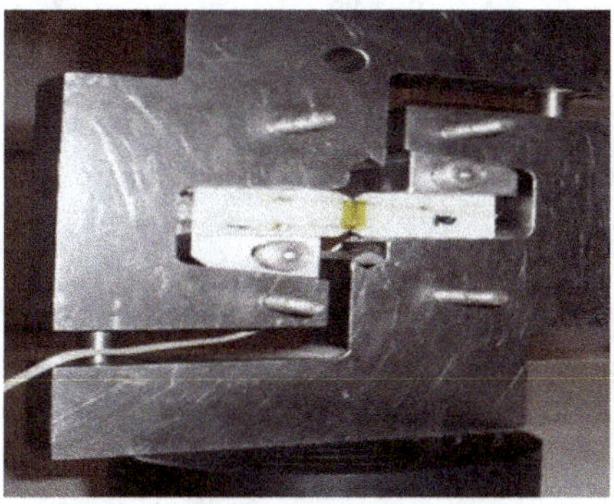

Fig. 2.25 Specimen with [0°/90°] orientation

Fig. 2.26 Specimen with [45°/0°/+45°/90°] orientation

The shear test on the specimen oriented at [0°/90°] resulted a shear strength of $\tau_{rf.90}$ = 54 MPa and a transverse modulus of elasticity of G_{90} = 7824.6MPa. Figure 2.27 shows the characteristic shear curve for the specimen at [0°/90°] with the correction of the specific shear at the transverse sensitivity. Figure 2.28 plots the approximation of the τ-γ curve in the elastic domain by a line whose equation is y=7824.6x-1.1847.

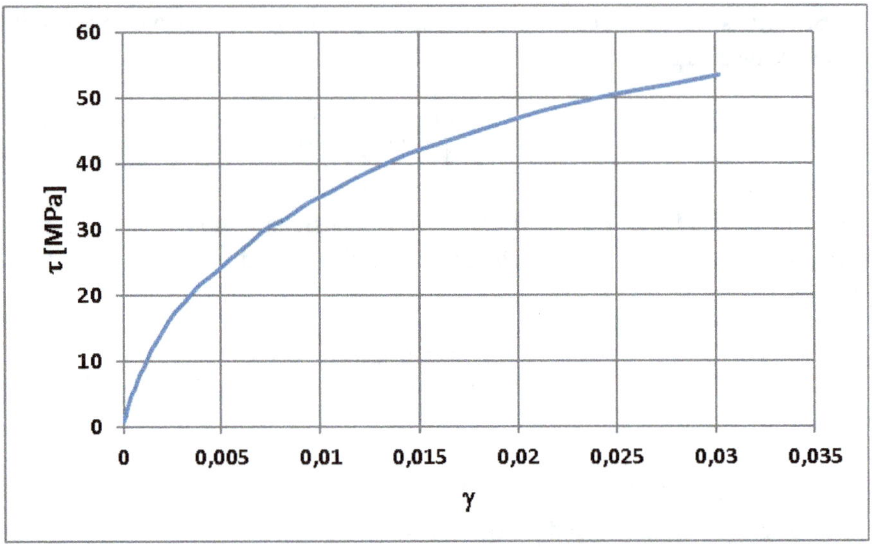

Fig. 2.27 Characteristic curve to pure share at [0°/90°]

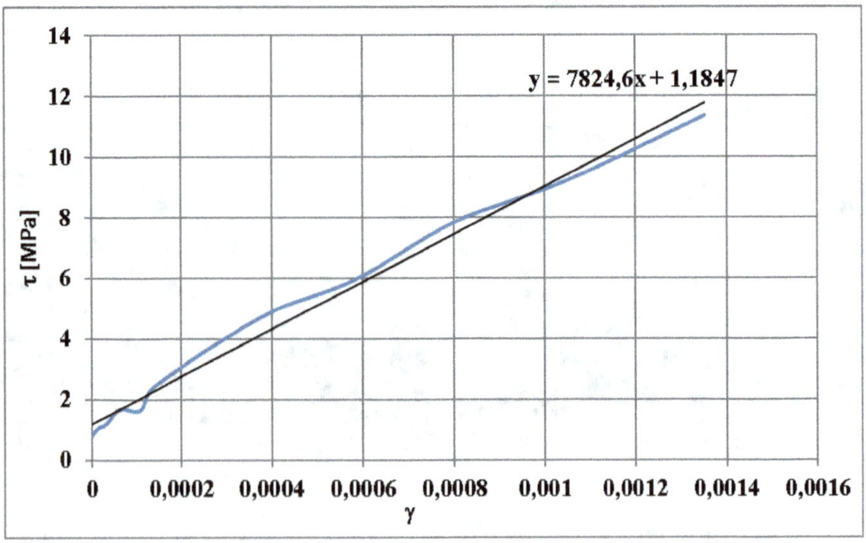

Fig. 2.28 Determination of in plane shear modulus G_{90}

The shear test on the specimen oriented at [-45°/0°/+45°/90°] resulted a shear strength of $\tau_{rf.45}$ = 88.5MPa and a transverse modulus of elasticity of G_{45}=8590 MPa. Figure 2.29 shows the characteristic shear curve for the specimen at [-45°/0°/+45°/90°] with the correction of the specific shear at the transverse sensitivity. Figure 2.30 plots the approximation of the τ-γ curve by a line whose equation is y=8590x-5,466.

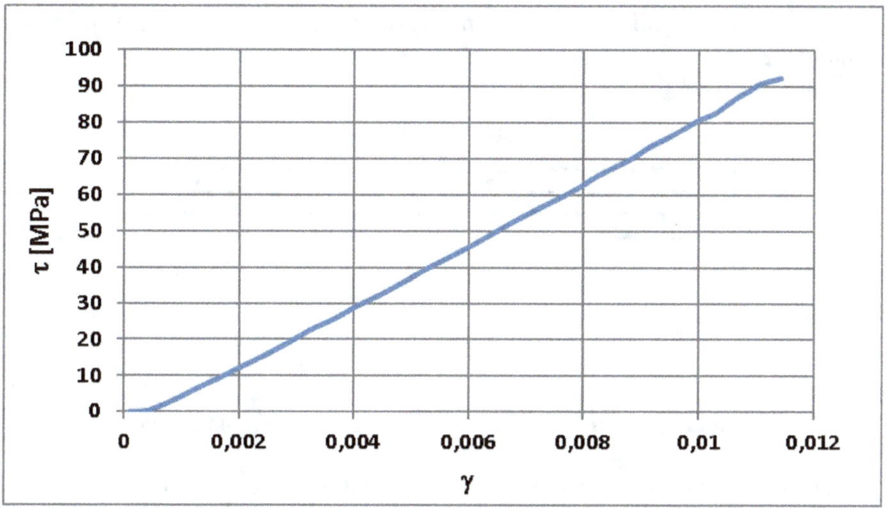

Fig. 2.29 Characteristic curve to pure share at [-45°/0°/+45°/90°]

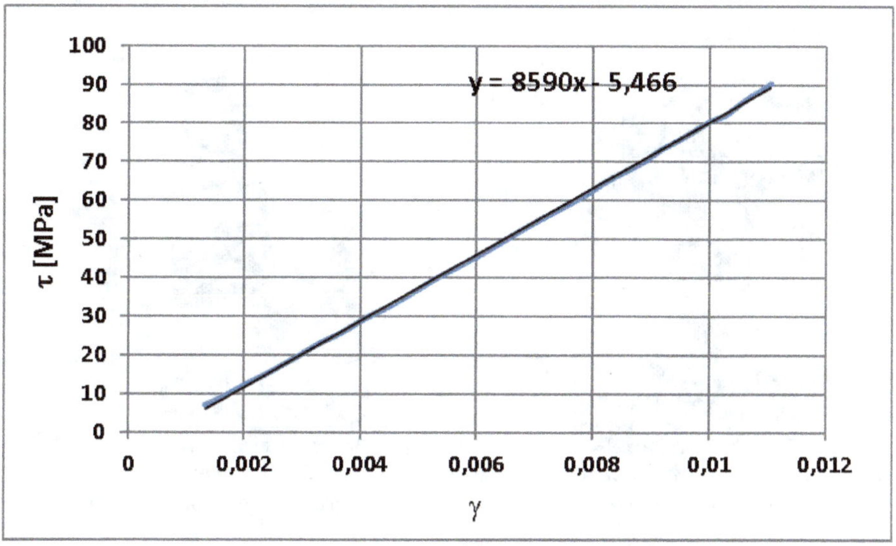

Fig. 2.30 Determination of in plane shear modulus G_{45}

2.6 Compression test

Compression tests were carried out on the same GFRP material, from the [0°/90°] fiber orientation plate, as presented in subchapter 2.2. Four specimens, cross-cut specimens, were used in the compression test, which followed the modified ASTM 695 standard (Boeing method), Figure 2.31. Figure 2.31 is denoted with 1- specimen; 2- tabs; 3- and 4 unidirectional strain gauge marks, glued on both sides in the calibration area.

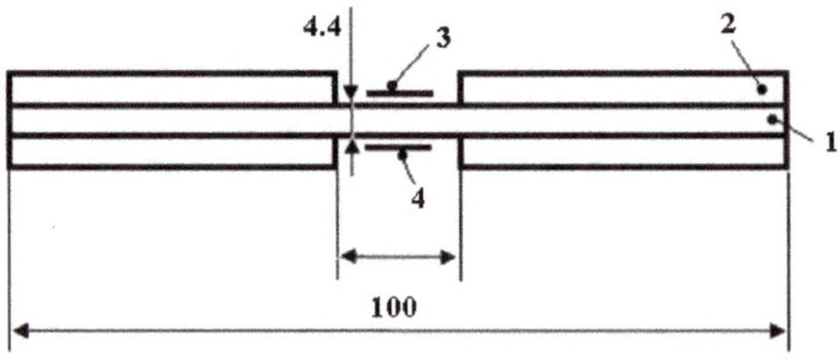

Fig. 2.31 GFRP specimen dimensions, according to ASTM 695

The compression tests were performed on the INSTRON 8801 universal testing machine (Figure 2.32a) at room temperature and a loading speed of 0.5mm/min. On the basis of the results obtained after breaking the specimens, stress-strain curves were plotted only for specimens TR 1, 2 and 3.

Fig. 2.32 a Test device mounted on the INSTRON 8801 testing machine

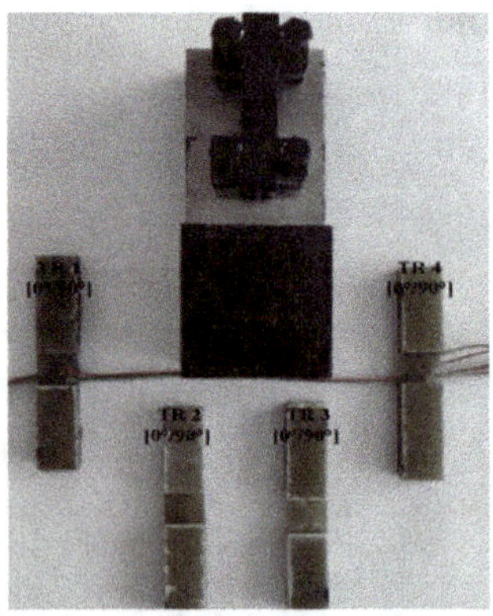

Fig. 2.32 b Compression test device and test specimen set

In order to obtain the Young's modulus, two unidirectional strain gauges of type EA-13-240LZ-120, manufactured by Micro-Measurements with resistance R=120Ω±0.3% and mark factor k_G=2.095±0.5%, were bonded on specimen TR 4.

Under these conditions, the testing took place in the elastic domain, with both strain gauges bonded in the longitudinal direction measuring the specific longitudinal strain ε_L. Figure 2.32b shows the device and the set of specimens for the first compression test method.

In order to obtain a better accuracy of the results, a statistical analysis of the data was performed and, in addition, for each failed sample, the mode of breakage was also followed, Figure 2.33.

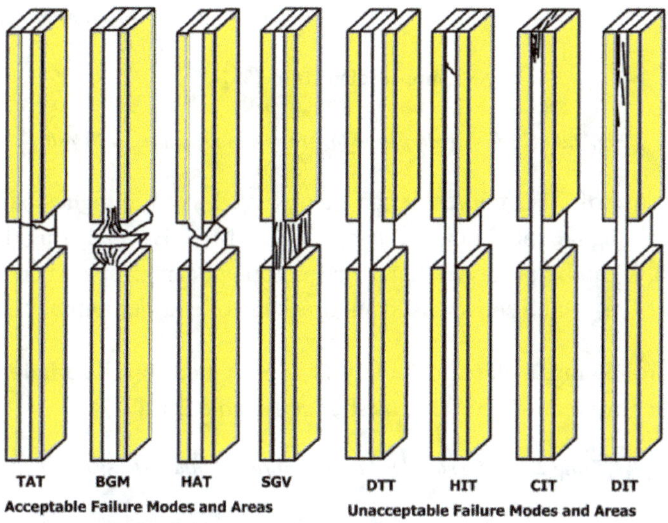

Fig. 2.33 Compression test failure codes/typical failure modes according to D 3410/D 3410M - 03

Figure 2.34 shows the overlaid characteristic curves for the 3 specimens cut transversely (in the weft direction) from the plate with fiber orientation at [0°/90°].

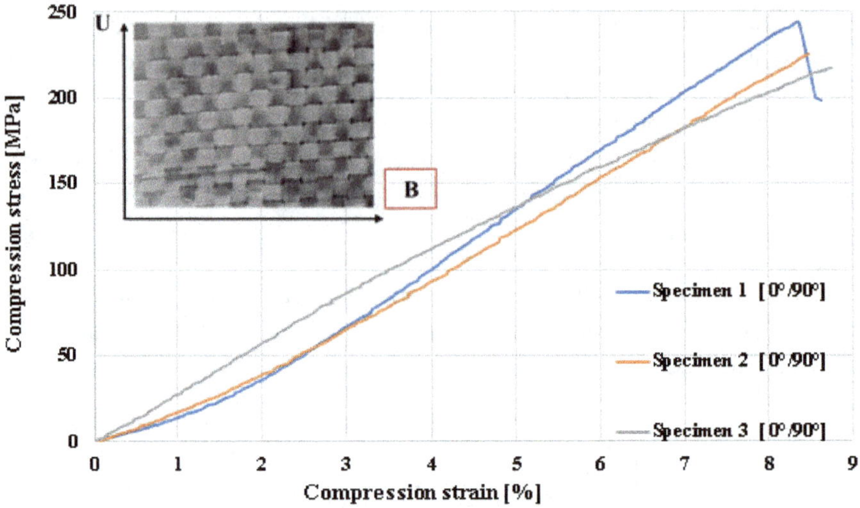

Fig. 2.34 Stress-strain curve for transverse specimens [0°/90°]

In the case of the 3 curves, it can be observed that the highest stress occurs at specimen 1, having the maximum value σ_r= 243.23 MPa. By analyzing the appearance of the 3 plots, it can be observed that a sharp drop appears at specimen 1 at the stress of about 240 MPa, which shows a sudden breakage of a larger number of fibers.

In the following, a statistical analysis is presented for all sample sets from the plate with fiber orientation [0°/90°], as shown in Table 2.12.

Table 2.12 The standard deviation for ultimate stress σ_{UTS}

Sample no.	σ_{UTS} [MPa]	$\bar{\sigma}_{UTS}$ [MPa]	Deviations from the mean $\sigma_r - \bar{\sigma}_r$ [MPa]	$(\sigma_r - \bar{\sigma}_r)^2$ [MPa]	Standard deviation S [MPa]	Coefficient of variation CV [%]
1	**243,23**	228,62	14,61	213,45	13,30	5,81
2	225,49		-3,13	8,94		
3	217,15		-11,47	131,56		
Σ=	685,67			353,95		

At the same time, the breakage of the specimens was analyzed and is presented in Table 2.13.

Tabel 2.13 Failure mode of specimens taken in transverse direction from the plate [0°/90°] according to ASTM D 3410/D 3410M - 03

Sample no.	Failure mode according to ASTM D3410/D 3410M-03 (Fig. 2.33)	First character	Second character	Third character
1	BGM	Brooming	Gage	Middle
2	HAT	Through	Thickness at tab	Top
3	BGM	Brooming	Gage	Middle

Figure 2.35 showed the mode of breakage for each specimen using a stereomicroscope. In all three cases, the degradation mechanism is acceptable according to the standard. The area of delamination failure brooming for specimens 1 and 3 occurred in the gaged area and for specimen 2 near the upper tab. The breakage location for specimens 1 and 3 occurred in the middle and for specimen 2 in the top.

a) TR1 b) TR2 c) TR3

Fig. 2.35 Failure mode of specimens taken in the transverse direction with fiber orientation at [0°/90°]

For the TR4 specimen, on which 2 unidirectional strain gauges were placed in the longitudinal direction, the Young's modulus was determined, Figure 2.36.

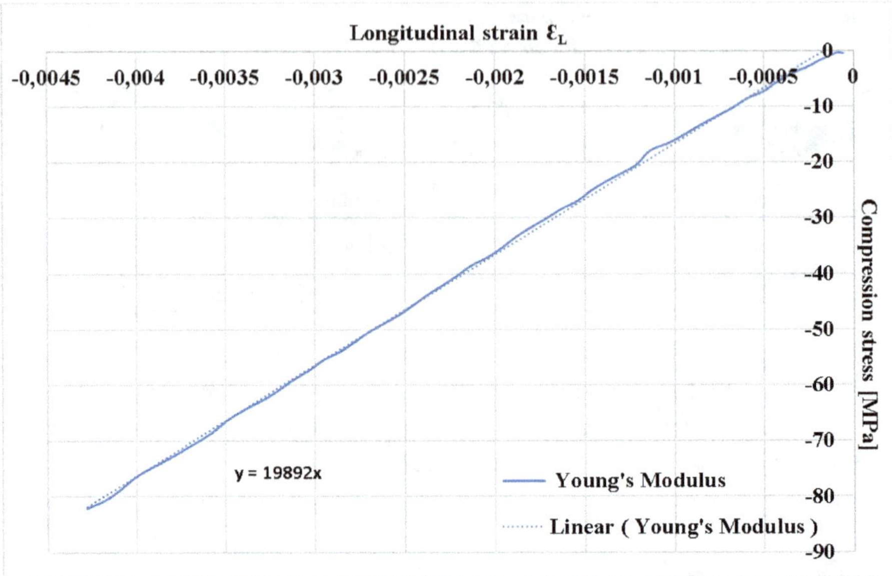

Fig. 2.36 Stress variation with longitudinal strain

In order to determine the Poisson's ratio of the studied GFRP, another compression test was performed. For the purpose of avoiding buckling, a special device has been developed in which four specimens of the same shape and size are tested at the same time.

The device clusters four specimens with $L \cdot l = 130 \times 25$mm, cut from 4.4mm thick GFRP plate, placed along the sides of a square, with an aluminum alloy cover and eight 0.5mm thick aluminum sheet spacers at each end, Figure 2.37, in which 1-front sample; 2-left sample; 3-left sample; 3-right sample; 4-Al-alloy Al-alloy cap; 5-Al-tin spacer; 6-T-type spacer T. The drawing of a spacer is shown in Figure 2.38.

On both sides of the frontal sample, two strain gauges type CEA-06-125WT-120, manufactured by Micro-Measurements, with resistance $R=120\Omega\pm0.35\%$ and gauge factor $k_G=2.025\pm0.5\%$, were bonded. The four rosette grids were coupled to a Vishay P3 tensiometer bridge (quarter-bridge mounted).

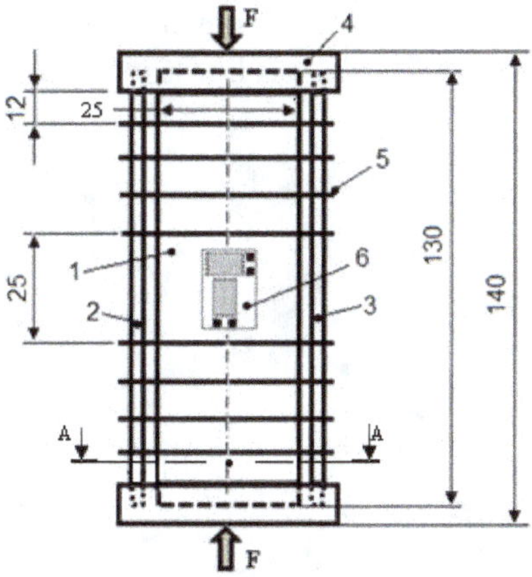

Fig. 2.37 Specimen for Poisson's ratio determination (back GFRP sample not shown)

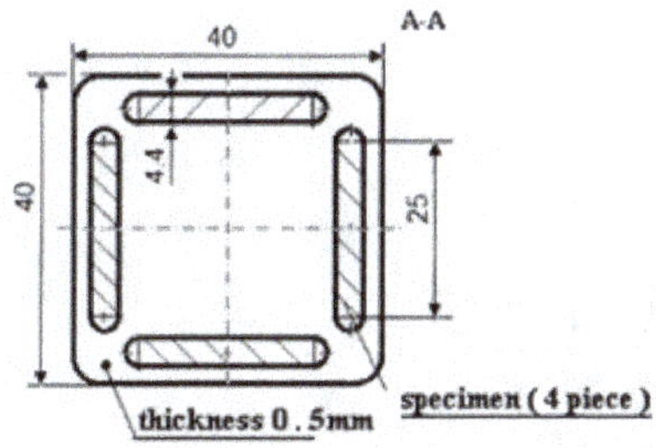

Fig. 2.38 Spacer

The specimen was placed between the platens of an Instron 8801 testing machine, Figure 2.39. The specimen was loaded in the elastic range at a rate of 0.3 mm/min up to a force of 82 MPa for a single specimen. Force-displacement diagrams of the machine traverse and output signals from the four strain gauge rosette grids were recorded during the test.

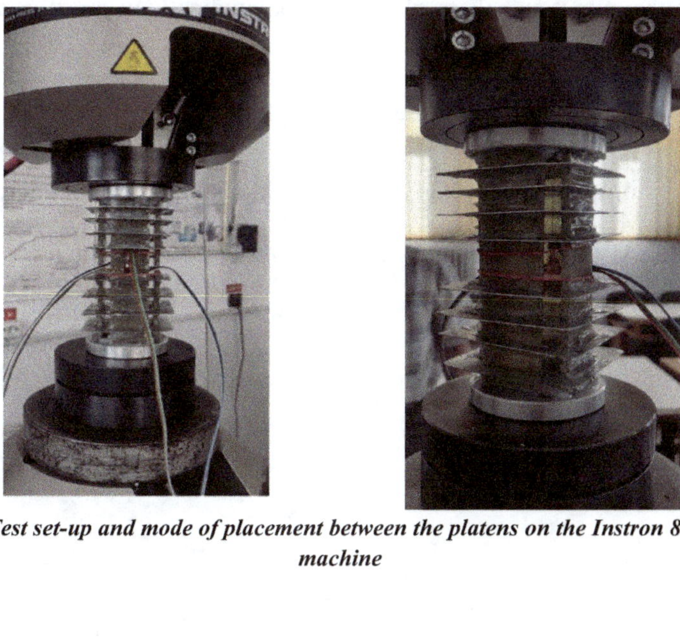

Fig. 2.39 Test set-up and mode of placement between the platens on the Instron 8801 testing machine

Figure 2.40 shows the stress-strain curve for a specimen subjected to compression. The stress was in the elastic range and the curve is linear in appearance.

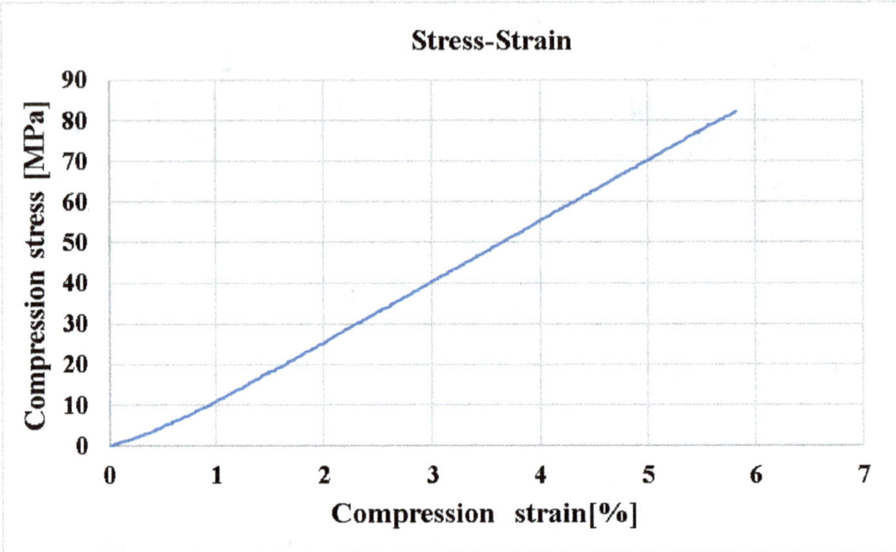

Fig. 2.40 The stress–strain curve from compression tests at [0°/90°]

Since the output signals from the two longitudinally oriented grids satisfy the condition for maximum force:

$$\frac{|\varepsilon_{LF}-\varepsilon_{LB}|}{|\varepsilon_{LF}+\varepsilon_{LB}|}\cdot 100=\frac{6991-6027}{6991+6027}\approx 7.4\% < 10\%,$$

and according to ASTM D 3410/3410M-03, buckling does not occur in the samples.

Figure 2.41 shows the output signals from the two strain gauges.

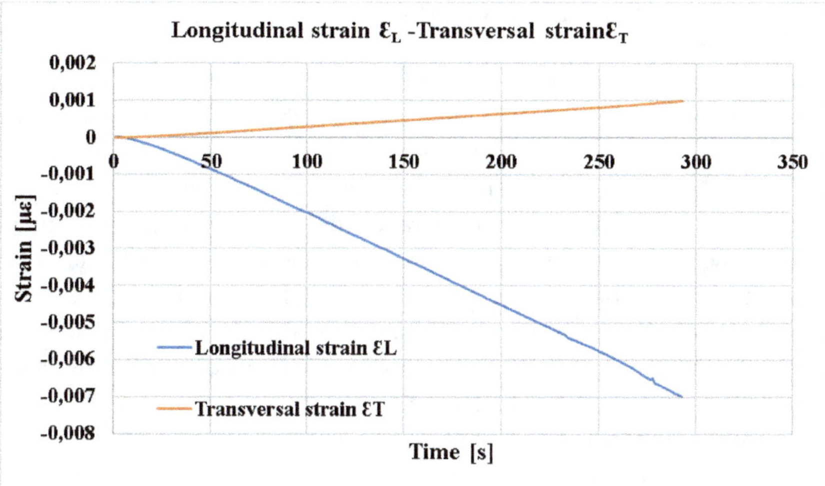

Fig. 2.41 The longitudinal strain ε_L- transversal strain ε_T curve

Figure 2.42 shows the transverse specific strain (ε_T) / longitudinal specific strain (ε_L) coordinate plots for the determination of the Poisson's ratio ν. For the studied GFRP, the Poisson's ratio ν=0.1427 was obtained in compression. For the tensile test of cross-cut specimens (notch direction) from the same GFRP slab, ν=0.1 was obtained (Morăraş C., et al., 2023)

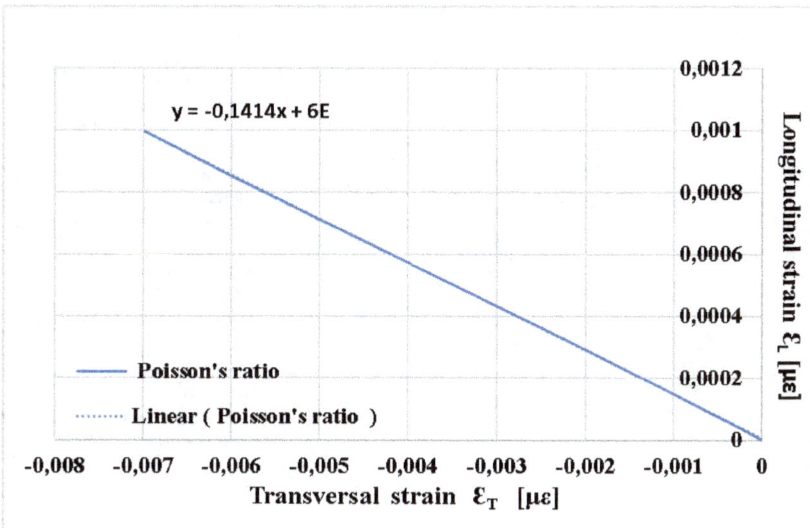

Fig. 2.42 Longitudinal strain-transversal strain

2.7 Influence of moisture on tensile properties of GFRP material

Tensile tests were carried out on specimens of the same GFRP material, from the [0°/90°] fiber orientation plate presented in subchapter 2.2. The specimens were immersed in regular water and seawater. The immersion was carried out for a certain number of days, namely 7, 14, 21, and 31 days. In addition, immersion in liquid nitrogen for 4 days at an initial temperature of -70°C was also performed. The immersion in liquid nitrogen was carried out with a view to the possible installation of wind turbines at high altitudes or in arctic areas. The results obtained from the tensile test, in terms of specific stresses and strains, were compared with each other and with those from the initial dry test. The immersions in regular water, salt water and liquid nitrogen of the GFRP samples were done to see the influence of humidity, existing salts or low temperature (Morăraş C., 2022). The aspects considered in order to see the influences of moisture and low temperature are: the shape of the characteristic stress - specific strain curve, the tensile strength value and the energy under the force-displacement curve which represents the energy that can be taken up by the sample. The shape and dimensions of the specimens conform to ASTM D3039.

The immersed specimens were tension tested on the INSTRON 8801 mechanical testing machine. Figure 2.43 shows the specific stress-strain curves for all immersion cases, including the sample immersed in nitrogen.

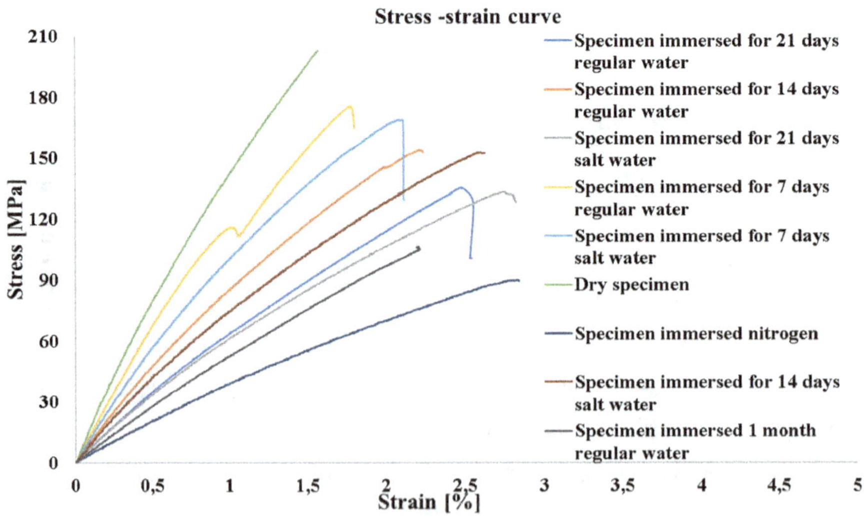

Fig. 2.43 Stress-strain curves for dry and immersed specimens

Analysis of these curves shows the following:

- Compared to the dry sample, all other samples have lower tensile strength;
- Compared to the dry sample, the deformations of the other samples are higher;
- With immersion in regular water, salt water or liquid nitrogen, it can be said that significant changes in strength characteristics occur.

Figure 2.44 shows the variation of the tensile strengths for each immersed specimen, and reveals the following:

- The highest ultimate tensile strength is for the dry, un-immersed sample;
- As the samples have been immersed for longer, the ultimate tensile strength decreases;
- The decrease is greater in samples immersed in regular water, the more days of immersion, the greater the decrease;
- As mentioned above, the sample immersed in liquid nitrogen was in full surface contact with nitrogen for four days, during which time the temperature did not rise above -700C. It was also found that we have here a decrease in tensile strength. This decrease is much greater than in samples immersed in water and removed after 7, 14, 21 and 31 days.

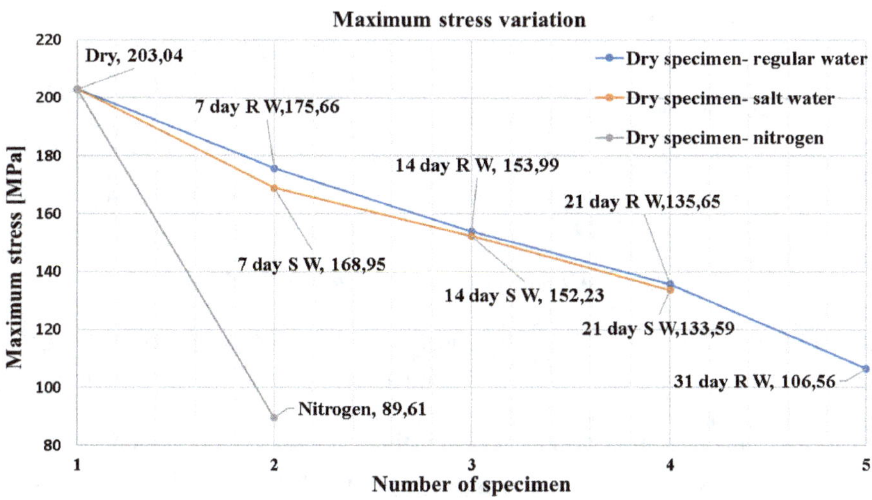

Fig. 2.44 Ultimate tensile stress variation versus the number of immersion days

Based on the records of the variation of the stress force F [N] with respect to the elongation Δl [m] of the specimen, the area under the corresponding curves is calculated. Given the shape of the curves, it can be said that the area determined by the variation F- Δl represents the mostly elastic strain energy. The elastic strain energy is an indicator of the ability of wind turbine blades to absorb wind energy through elastic deformation. Observing Figure 2.45, the following conclusions can be drawn:

- Dry sample accumulates the most energy;
- As the number of days of immersion increases, the energy decreases;
- Significant energy decreases are found in the 31-day immersed sample and liquid nitrogen;
- The variation in energy for 7, 14 and 21 days for the two types of immersion is not very marked;

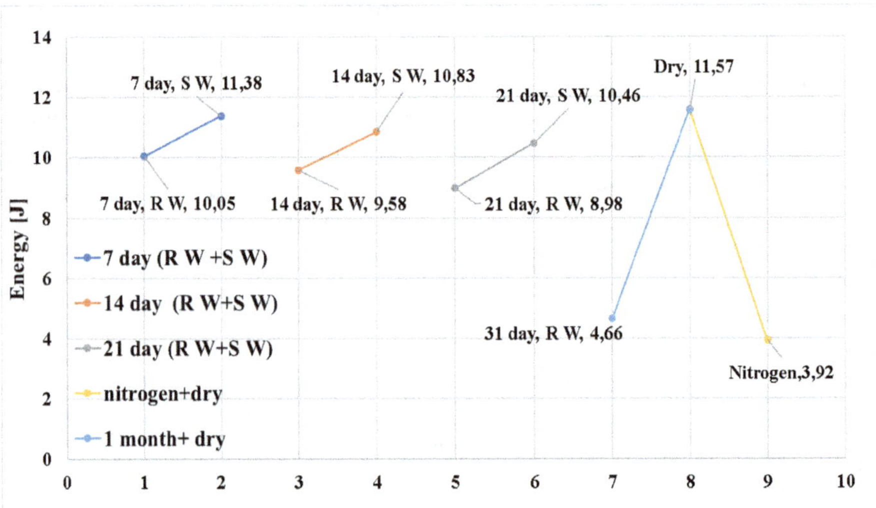

Fig. 2.45 Variation of energy with number of days

In Figures 2.46 and 2.47 the dry test specimen was taken as a reference and comparisons were made according to the immersion solution and the number of days:

- For specimens immersed in salt water, there is a slow decrease in energy;
- In the case of specimens immersed in normal water a sharp decrease can be observed between 21 and 31 days. A possible reason may be the destruction of the fiber-matrix interface;

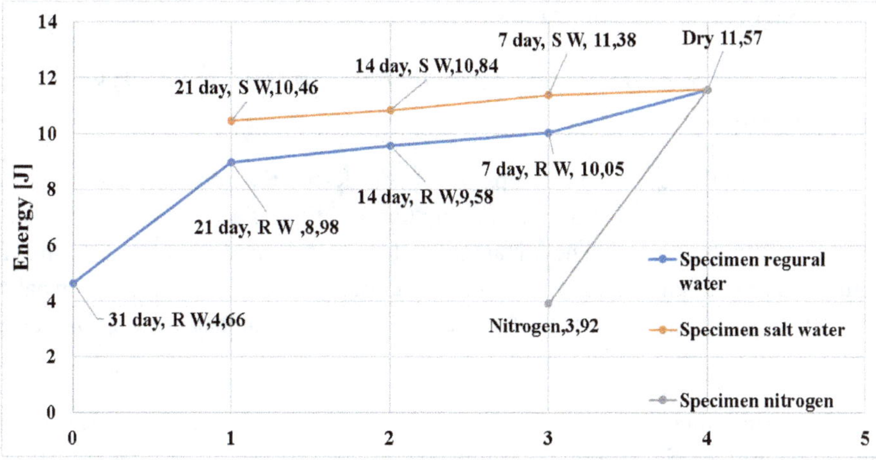

Fig. 2.46 Variation of energy with number of days

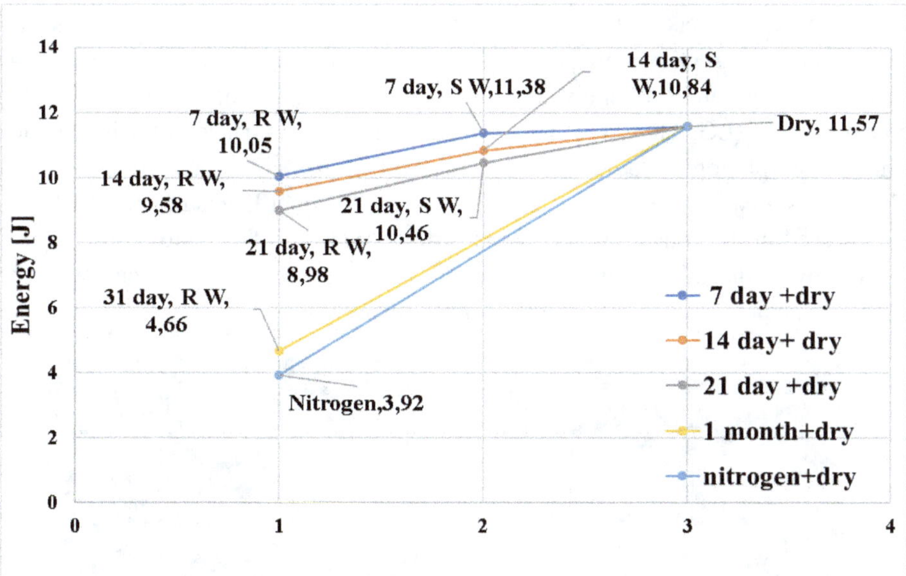

Fig. 2.47 Energy variation with dry sample compared to other immersion days

2.8 Structural analysis of GFRP composite material

2.8.1 SEM on GFRP material with fiber orientation at [0°/90°]- tensile specimens

Scanning Electron Microscopy is a non-destructive analysis technique that allows the microstructural and morphologic highlighting of a wide variety of metallic and non-metallic materials. The advantages of using this analysis technique are: high resolution images, ease of preparation of samples for analysis, and the wealth of information collected (porosity, identification of overall/point chemical composition, etc.).

Electron microscopy analysis, using the SEM Quanta 2003D electron microscope, was carried out on the samples with the highest tensile strengths. In order to improve the electrical conductivity, a SEM COATER Luxor Au - CT-2201-0144 was used, which realized a 7 nm gold layer on the surface and in section. The aim was to highlight clear aspects of fiber breaks and characteristic areas. SEM images of the surface and cross-sectional area of the GFRP base material are shown in Figure 2.48a, Figure 2.48b. Figure 2.49 shows the cross-sectional images of the specimens in the three directions (LG, TR and 45°). Figure 2.48a shows the surface of the base material from the morphological point of view. The surface has a compact structure and no microcracks are observed. Figure 2.48b shows the 10 fiber layers of the GFRP cross-section used in the construction of wind turbine blades. In the same figure, the SEM image showed that fiber pullout is a primary damage mechanism, which was evident when the specimens were subjected to higher stresses.

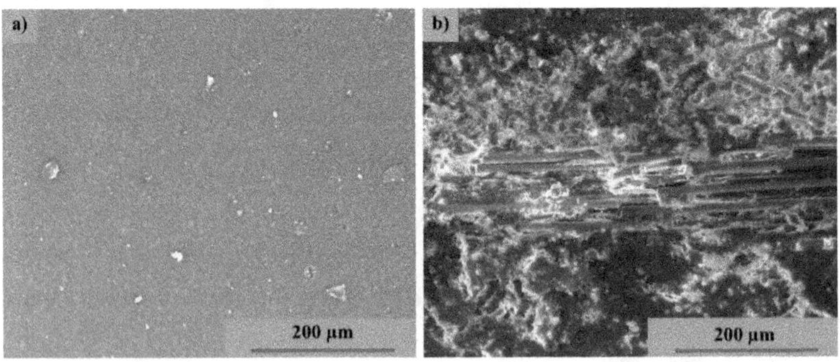

Fig. 2.48 SEM images of the base material—GFRP: a) surface image and b) cross-section image

Figure 2.49. shows images of the cross-section from which the samples were obtained. Thus, Figure 2.49a shows a fiber break in the transverse direction, while

Figure 2.49b highlights a fiber break at 45°, these two breaks having a major damage appearance. Figure 2.49c shows a delamination at the intermediate layers, with the breaks showing moderate behavior. The failure mode, in all three specimen cases, represents a typical brittle fracture characteristic. GFRP materials are mainly composed of the 2 basic elements, C and O. The material with lighter colored surfaces is attributed to O-rich areas, opposite to darker C-enriched surfaces.

Fig. 2.49 Cross-section SEM images of GFRP specimens in the 3 stress directions: a) transversal, b) oriented at diagonal and c) longitudinal

Figures 2.50-2.52 show the surface morphology of the samples examined in all three stress directions. In all three examples, a comparable brittle-type fracture process is observed in the fiber-resin structure. The different magnification strengths demonstrate the fracture behavior and appearance of the polymer fibers pulled out at the interface with the resin. Figure 2.50 highlights a more compact and evenly dispersed pattern of these fibers, while Figures 2.51 and 2.52 demonstrate more significant dislocation.

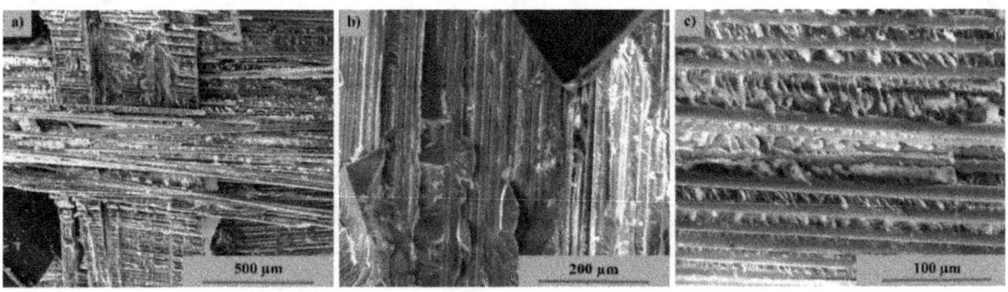

Fig. 2.50 Surface SEM images of transversal [0°/90°] GFRP sample: a) 200X, b) 500X and c) 1000X

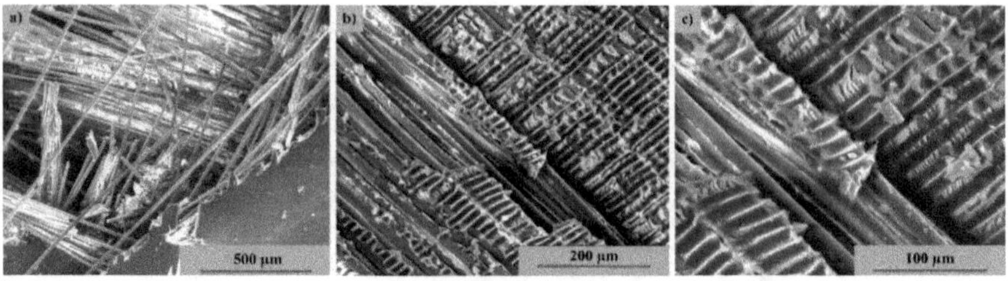

Fig. 2.51 Surface SEM images of diagonal [0°/90°] GFRP sample: a) 200X, b) 500X and c) 1000X

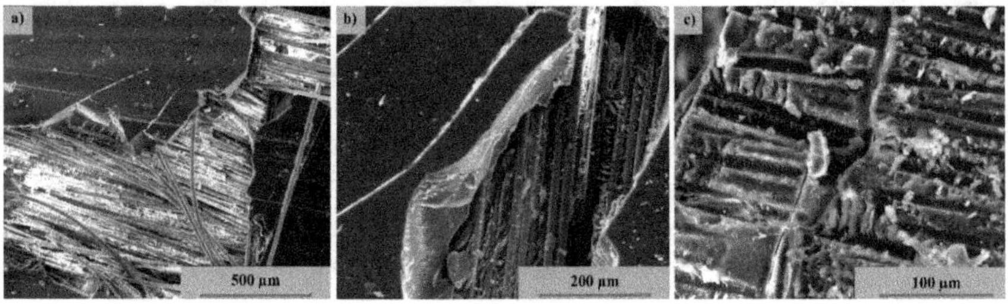

Fig. 2.52 Surface SEM images of longitudinal [0°/90°] GFRP sample: a) 200X, b) 500X and c) 1000X

2.8.2 SEM on GFRP material with fiber orientation at [-45°/0°/+45°/90°]-tensile specimens

Figure 2.53 shows SEM images of the surface and cross-section of the GFRP base material. Figure 2.53a shows the surface of the base material morphologically. This surface shows a compact area, highlighting the presence of several small microcracks. Figure 2.53b shows the structure of the composite material consisting of the base matrix and the epoxy resin - EPIKOTE. This is a low-viscosity epoxy resin consisting of bisphenol, epichlorohydrin and containing an added proportion of mono-epoxide alcohol, i.e. OH, CH_3, Cl and CO groups.

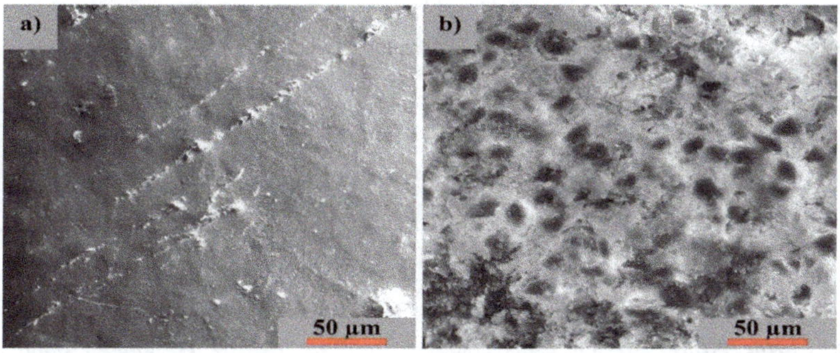

Fig. 2.53 SEM images of the base material GFRP: a) surface image and b) cross-section image

Figure 2.54 shows the cross-sectional images of the tested specimens in the (TR and LG) direction. Figure 2.54a shows fiber breakage in the transverse direction, while Figure 2.54c highlights fiber breakage in the longitudinal direction. These two types of breaks have a high damage appearance. Figure 2.54d. shows minor delamination at the intermediate layers, with failure showing moderate behavior.

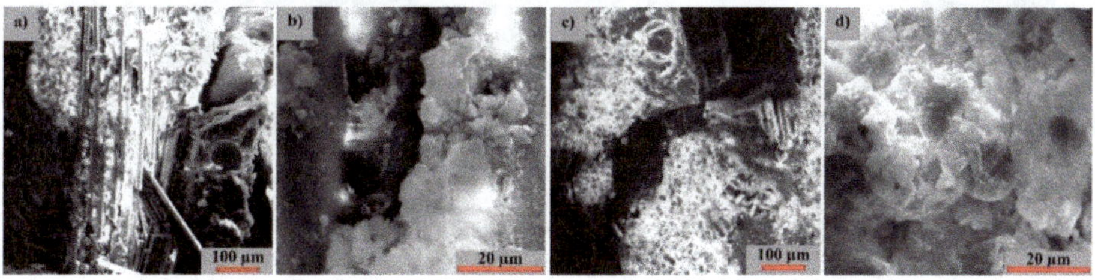

Fig. 2.54 Cross-section SEM image of GFRP specimens in the 2 stress directions: a) 200 X (TR), b) 2000 X (TR), c) 200 X (LG) and d) 2000 X (LG)

Figures 2.55 and 2.56 show the surface morphology of the specimens studied in the two directions of loading, transverse and longitudinal. In each of the two cases, both the matrix and the fiber exhibit a similar brittle-type fracture mechanism. Figure 2.55 illustrates a more compact and uniform morphological pattern of these fibers, while Figures 2.56a and b show a more pronounced dislocation. Thus, matrix crack initiation was around the porosities and its propagation was along the fibers oriented at 45° to the mechanical loading direction.

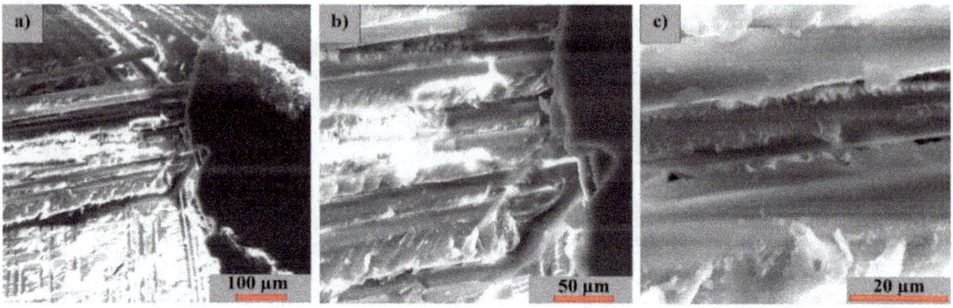

Fig. 2.55 Surface SEM images of transversal [-45°/0°/+45°/90°] GFRP sample: a) 200 X, b) 500 X and c) 2000 X

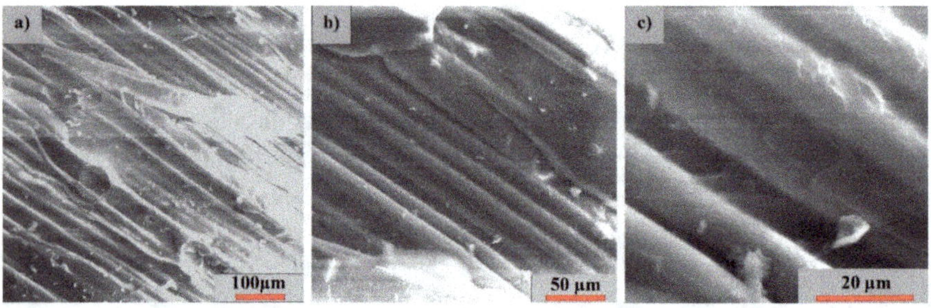

Fig. 2.56 Surface SEM images of longitudinal [-45°/0°/+45°/90°] GFRP sample: a) 200 X, b) 500 X and c) 2000 X

2.8.3. SEM on GFRP material with orientation [0°/90°]- compression specimens

To highlight the morphology of the experimental specimens, an SEM microscope was used to show the directions of the reinforcing fibers, the mode of fracture in the area of interest of the GFRP composite material. The fracture zone on the surface and in the cross section of the material was studied for specimens TR1, 2 and 3. To improve the electrical conductivity, a Luxor Au - CT-2201-0144 SEM COATER Luxor Au - CT-2201-0144 was used, which realized a 7 nm gold layer on the surface and in cross-section. For Figures 2.57-2.59, a TESCAN Vega SEM microscope Vega (Brno, Czech Republic) was used.

Figures 2.57-2.59 show morphological characteristics at different magnification levels, on the surface, of the specimens that were subjected to compression tests. In Figures 2.57-2.59, visual examination of the specimen reveals a uniform alignment of fibers within the matrix, indicating a dense bond between the fibers and the matrix. The surface shows an adhesion between the fibers and the matrix due to the significant resin content, thus reducing the occurrence of voids, Figures 2.57a and 2.59a. The composition of the material results in the formation of a robust substance that exhibits increased load carrying capacities, thus indicating a significant degree of impact resistance.

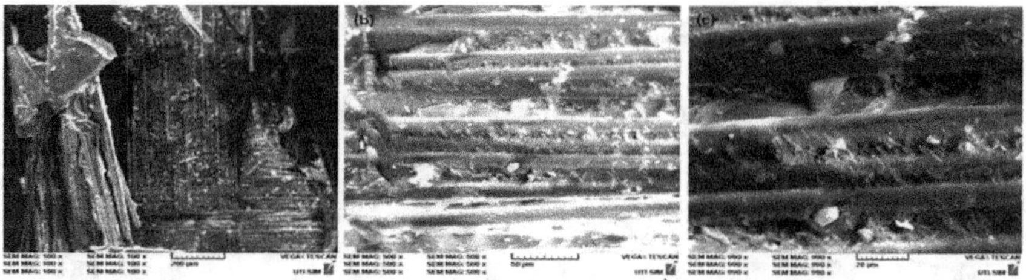

Fig. 2.57 SEM images of the surface material – GFRP specimen TR1: a) 100X, b) 500X and c)1000X

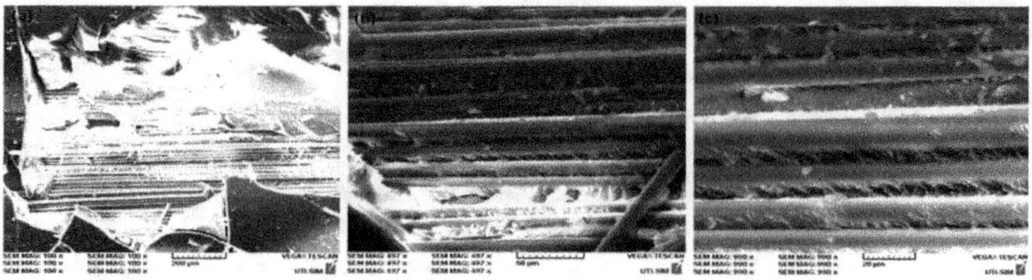

Fig. 2.58 SEM images of the surface material – GFRP specimen TR2: a) 100X, b) 500X and c) 1000X

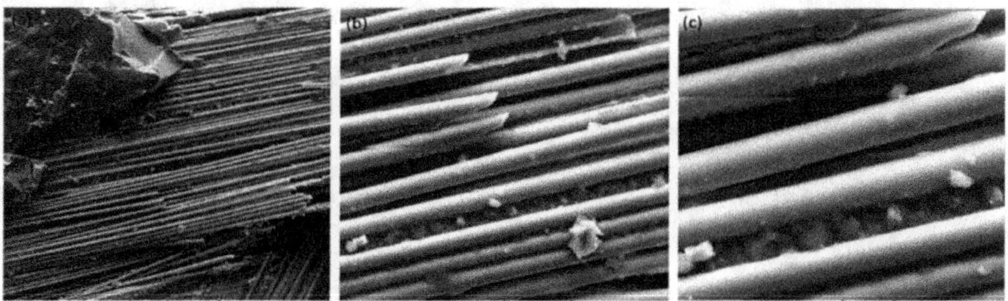

Fig. 2.59 SEM images of the surface material – GFRP specimen TR3: a) 100X, b) 500X and c) 1000X

Cross-sectional SEM images of specimens subjected to the compression test were shown in Figures 2.60-2.62. An obvious separation of the glass fiber reinforced fiber reinforced polymer (GFRP) fibers from the supporting matrix can be observed, probably attributed to the manufacturing process used in the fabrication of the original composite board (see Figure 2.60a). Fiber separation on contact with the matrix was observed to occur over an angular range of 0°-90°.

Cracks can be observed along the base matrix, as shown in Figures 2.61a and 2.62a, and can be attributed to the presence of spherical voids within the resin. Among the elements that induce crack propagation and can lead to the formation of multiple cracks are porosity, fiber interface, and particle interaction. Microcracks have the potential to propagate between two distinct regions: the interface between the fiber and the matrix and within the matrix itself.

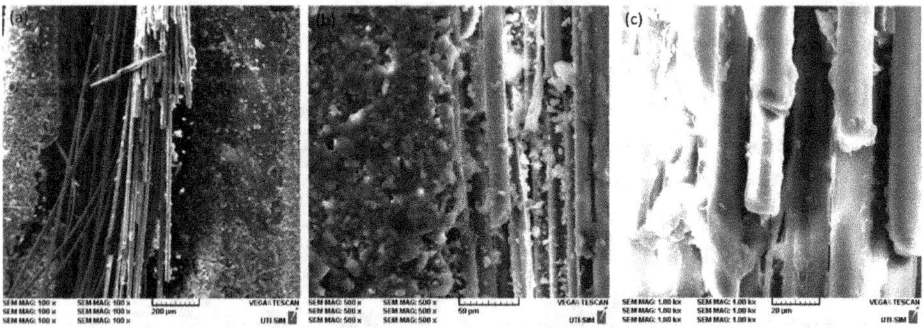

Fig. 2.60 Cross-section SEM images of GFRP specimen TR1: a) 100X, b) 500X and c) 1000X

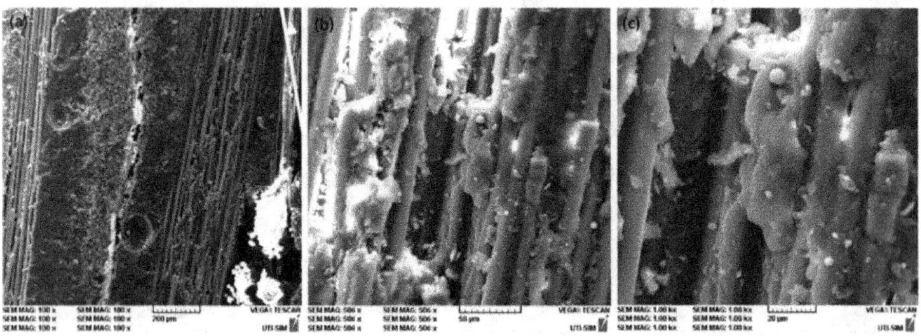

Fig. 2.61 Cross-section SEM images of GFRP specimen TR2: a) 100X, b) 500X and c) 1000X

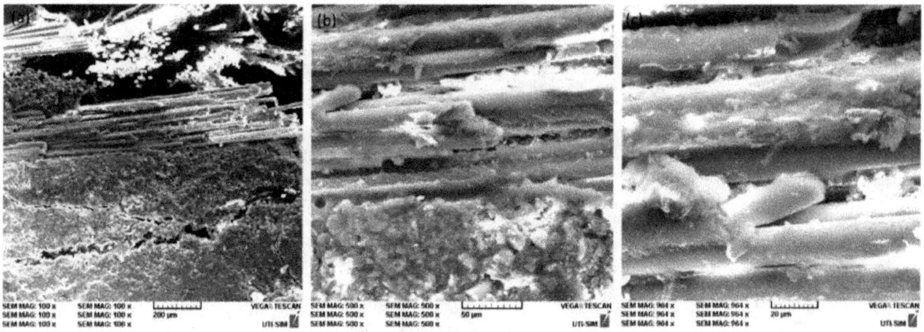

Fig. 2.62 Cross-section SEM images of GFRP specimen TR3: a) 100X, b) 500X and c) 1000X

2.9 X-ray diffraction analysis

X-ray diffraction is an additional method of non-destructive analysis, complementary to scanning electron microscopy. It is a versatile and efficient analysis to reveal the phasic constituents and the corresponding lattice parameters. The advantages of X-ray diffraction are the study of crystallographic bonding of compounds, residual stress and texture determinations. In the experimental analysis, the identification of the amorphous structure by X-ray diffraction was realized. Figure 2.63 comparatively highlights the X-ray diffractograms for the 5 types of GFRP material specimens, where a relatively amorphous structure is observed, due to the "hump" appearance of the main peak, identified approximately at the angle 2Theta=20°.

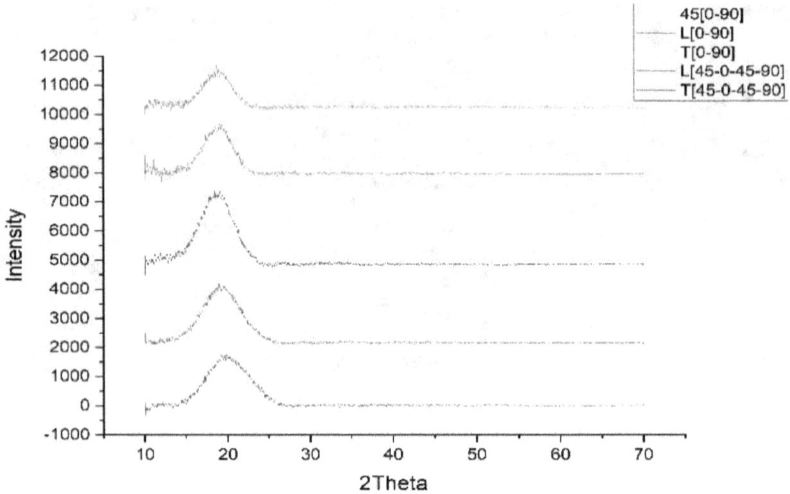

Fig. 2.63 X-ray diffractogram of experimental samples

However, the materials also exhibited crystallization peaks in the range 2Theta=10°-20°, which characterize a possible surface crystallization of the observed glass fiber.

2.10 General conclusions to Chapter 2

1. The tensile tests demonstrate an orthotropic behavior of the material, due to the different shear stresses occurring in different directions on the two GFRP plates. The specimens that were cut transversely from the [0°/90°] plate had the highest ultimate tensile strength. This is mainly caused by the alignment of the reinforcing fibers (reinforcement);
2. The results obtained from tensile tests of specimens cut at 45° of the plate [0°/90°] showed lower tensile strength values than those obtained in the other two directions. On the other hand, the elongation and deformation of the specimens cut at 45° were much higher than the other two directions;
3. The results obtained in compression tests on specimens cut in the transverse direction from the plate oriented at [0°/90°] are in good agreement with those in tensile. The material constants ν and E are approximately close;
4. A decrease in (elastic) strain energy is observed in all cases of immersed specimens. Both tensile strength and strain energy are lower in the case of

immersion in salt water. Nitrogen immersion also leads to decreases in both tensile strength and strain energy. These decreases are not as large as in the case of immersion in water (normal and salt), perhaps because the immersion time was also much shorter. The only parameter that has higher values after immersion is the strain, in all cases;

5. The research carried out contributes to a more accurate description of the material studied. Other types of composite materials can be described using similar methods.

CHAPTER 3. TESTS PERFORMED ON GFRP PRE-FABRICATED CT SAMPLES – FATIGUE CRACK GROWTH

3.1 Theoretical considerations for determining fatigue crack growth rate

For any part that may contain defects as early as during the technological production of the material or is susceptible to accumulate defects in service, it is important to determine the characteristic parameter of fracture mechanics called crack propagation rate. Research on how crack propagation occurs in GFRP composites is not very extensive. Crack propagation in these materials has several causes, as shown âabove. Here we mention a few:

- Air voids occurring during the manufacturing process;
- Delamination, which involves separation of layers of reinforcement matrix, either from each other or from the reinforcement;
- Cracks or fissures that may occur as a result of exposure to repeated stress or external factors;
- Chemical or erosive attack;
- Various manufacturing defects.

The propagation of cracks in these defects can lead to partial or total destruction of the component, with no possibility of further operation. The problem that arises if a defect is introduced into a component made of GFRP is whether it will last for a certain period of time in service. One of the characteristics provided by Fracture Mechanics that can provide an answer to the previous question is the crack propagation rate, da/dN [mm/cycle], which is determined with specific equipment and procedures, where **a** is the length of the propagated crack and **N** is the number of stress cycles that have led to crack propagation. Understandably, the stress used to determine da/dN is the fatigue stress. The resistance of a material to further propagation of an already initiated crack is provided by the a-N curve, Figure 3.1, where **a** is the length of the propagated crack and **N** is the number of cycles given by the fatigue stress, Figure 3.2 (Goanță V., 2006). The plotting of the a-N variation curve is a basic problem in Fracture Mechanics because it can be used to estimate the lifetime of a strength element or a structure in which a crack has initiated. Figure 3.1 shows that N_{cr} cycles are necessary to propagate a crack from the initial detected length a_0 to the critical unstable propagation length a_{cr}, which can be determined for each material on the basis of tests on CT (Compact

Tension) samples pre-fitted by fatigue. If defectoscopic methods (e.g. using the liquid penetrant method) show that the crack is already of length a_1, the remaining crack lifetime for the crack to reach the critical length from which propagation will still be unstable can be predicted from previous laboratory determinations.

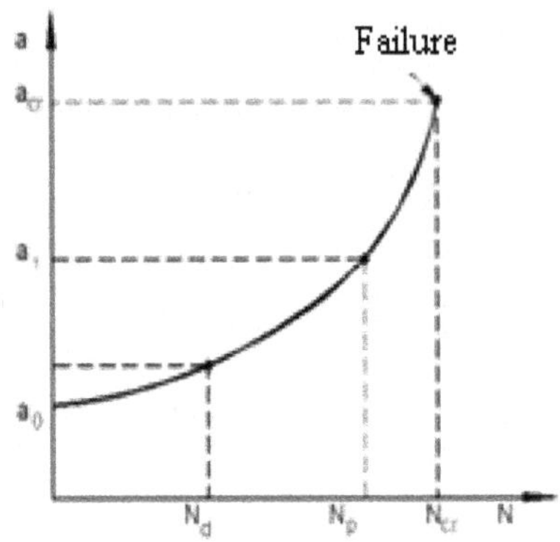

Fig. 3.1 Variation of crack length during cyclic loading

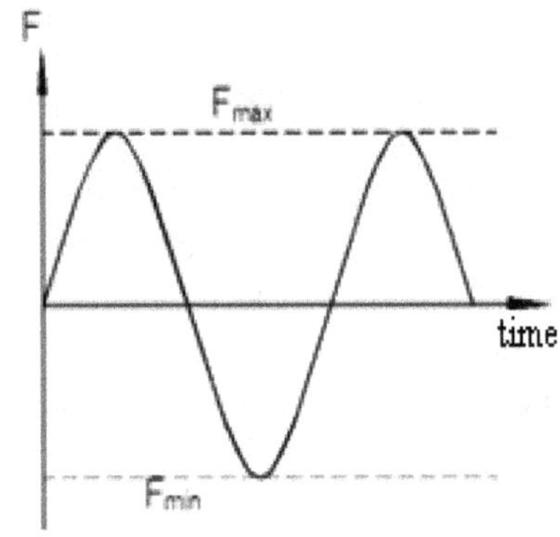

Fig. 3.2 Alternating symmetric fatigue stress cycle

For different crack lengths, the speed of crack propagation can be obtained by calculating the slope of the *a-N* diagram. The fatigue crack growth rate $\frac{da}{dN}$ depends on the initial crack length and the applied stress level or amplitude, which are involved in the expression of the stress intensity factor, K. Typically, with increasing initial crack length and increasing stress level, the fatigue crack propagation velocity increases. As a result, the fatigue crack growth rate can be correlated with the variation of the stress intensity rage ΔK, $\frac{da}{dN} = f(\Delta K)$. One of the computational models for fatigue crack growth rate was proposed by (P.C. Paris, 1965) who found that, for a fairly wide range of fatigue crack propagation velocity values, this quantity can be calculated with the relation:

$$\frac{da}{dN} = C(\Delta K)^n \tag{3.1}$$

where $\Delta K = K_{max} - K_{min} = K(\sigma_{max}) - K(\sigma_{min})$ is the variation of the stress intensity factor for crack propagation *mode I* and *C* and *n* are constants which depend not only on the material, but also on the stress conditions, for example the cycle asymmetry coefficient. Paris' law is one of the most widely used equation in fatigue crack propagation analysis.

Being empirically established, it leads to results close to reality for a relatively large number of materials under different loading conditions.

3.2 Test program description

There are several Fracture Mechanics standards operating in the field of GFRP materials. Thus, ASTM D6115-97(2019) is available, which presents a test method for fatigue test *mode I* for the occurrence of fatigue delamination of unidirectional fiber-reinforced polymer matrix unidirectional composite materials. ASTM E1922 is intended for the determination of translaminar fracture toughness of layered polymer matrix composites. Since there is not yet a standard for the determination of crack propagation rate for GFRP materials, ASTM E647-23a, a test method for measuring the rate of fatigue crack growth, has been adopted for the present determinations. This standard does not specify to which type of material it is addressed. The test method prescribed in the above-mentioned standard provides for cyclic loading of notched specimens that have been pre-fissured by fatigue. The crack length is measured as a function of the number of fatigue cycles performed up to that point and the data is subjected to numerical analysis to determine the rate of crack propagation. The fatigue crack growth rate is expressed in terms of the stress intensity rage, (ΔK), which is calculated based on the linear-elastic domain analysis. Materials that can be tested by this method are not limited in thickness or strength, provided that the specimens are thick enough to prevent buckling and that during the test, the area slightly beyond the crack tip remains within the elastic range. The test method defines a wide range of sizes, with proportional sizes, but the size may be modified according to the yield strength and the applied load.

The tensile specimen or CT (compact tension) specimen has a single lateral notch in the extension of which a fatigue crack is induced to be as close in appearance as possible to the actual crack. The general configuration and dimensions of this specimen are shown in Figure 3.3.

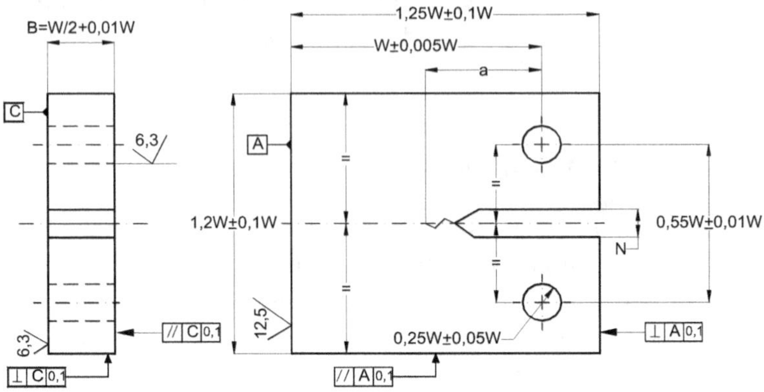

Fig. 3.3 Specimen of tensile test – CT (compact tension)

The results are expressed in terms of the crack tip stress intensity factor, ΔK, which is defined by linear elastic theory. The program/software "da/dN" contains different ways to determine fatigue crack propagation. Crack propagation is usually determined visually, but there are other methods such as potential difference, determination of discharge compliance, etc.

The "da/dN" software implemented on the PC running the testing machine is fully compliant with ASTM E647 (Standard Test Method for Measurement of Fatigue Crack Growth Rates). There are a number of features to be entered in the main menu of the software: material data, specimen configuration and geometry, specimen dimensions, strain gauge location, etc. The software uses several modalities concerning the determination of the crack propagated length in real time. Among these, there is the visual mode with instantaneous measurement and menu entry, as well as the mode using specimen compliance. The relationship between compliance and crack length has been analytically determined for a number of standard specimens, including the compact specimen (CT). This relationship, together with other helpful relationships have been implemented in the standard.

These relations are usually expressed in dimensionless terms, for compliance having the relation:

$$U = \frac{1}{\sqrt{\frac{E \cdot v \cdot B_{eff}}{P}} + 1} \tag{3.2}$$

where B_{eff} is the effective sample thickness and is given by the relation:

$$B_{eff} = B - \frac{(B - B_{net})^2}{B} \tag{3.3}$$

and the normalized crack length, a/W is defined by the relation:

$$\frac{a}{W} = \sum_{i=0}^{5}(C_i \cdot U^i) \tag{3.4}$$

where:

- E- Young's modulus;
- v - the displacement between the points where the extensometer is attached;
- B_{eff} - the effective thickness of the sample;
- P - force between measuring points;
- a - crack length;
- W - sample width, see Figure 3.3;
- $C_{0...5}$ or C_i being the compliance coefficients.

For the compact specimen (CT), the coefficients C_i are given for both the front face and the load line, depending on where the strain measurement is taken – either using an extensometer for the front face or based on the crosshead displacement at the points of force application.

The data obtained from the fatigue crack growth rate test is used to determine the suitability of the tested material for the applications to which it will be subjected in service. The information obtained from such a test is important in the selection of the appropriate material for components and structures where fatigue cracking may be critical to the service life of that component or structure.

3.3 Material used

CT (Compact Tension) samples from two 10 mm thick GFRP (Glass Fiber Reinforced Plastics) plates were used for fatigue crack growth rate and toughness determinations. Aspects concerning the characteristics of the EPIKOTE MGS LR 385 resin and the processing of the composite material were presented in chapter 2.

The tension specimen or CT (Compact Tension) specimen has a single lateral notch in the extension of which a fatigue crack is induced to be as close as possible to the actual crack in appearance. The general configuration and dimensions of this specimen are shown in Figure 3.3. The specimen used in the experiments was of thickness B = 10 mm, all other dimensions being derived from this.

The first composite board is composed of 12 layers, Figure 3.4, which has fiber orientation at [0° /90°]. Fifteen specimens taken in three different directions (5 transverse (TR) in the direction of the weft, 5 longitudinal (LG) in the direction of the warp and 5 diagonal at 45°) were cut from the plate, Figure 3.4.

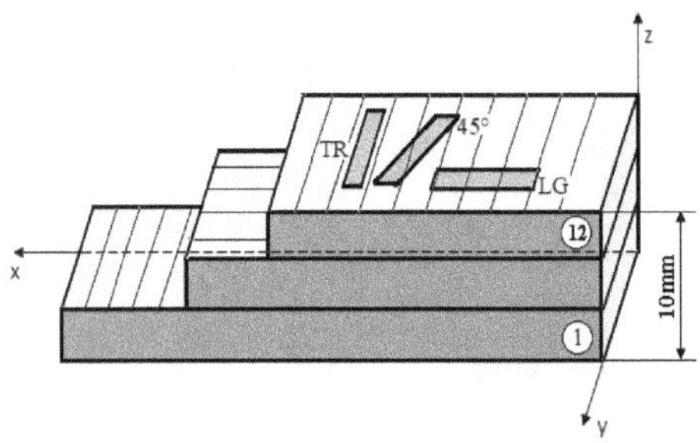

Fig. 3.4 Cutting directions of the GFRP sample [0°/90°]

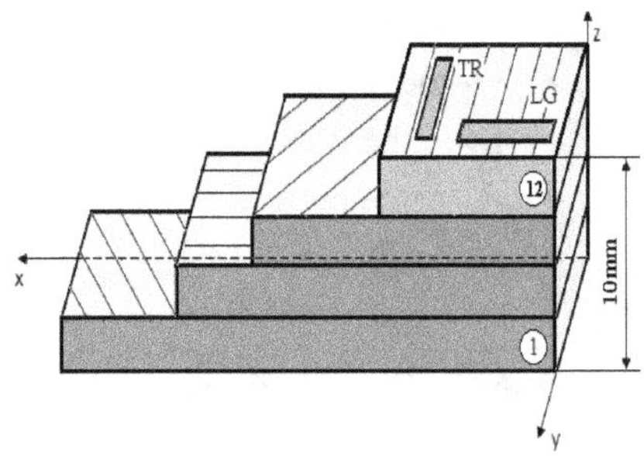

Fig. 3.5 Cutting directions of the GFRP sample GFRP [-45°/0°/+45°/90°]

Five specimens of each category were made, but some of the specimens broke during fatigue pre-cracking and could not be used.

The second composite board has also 12 layers like the first one, but the fiber orientation is at [-45°/0°/+45°/90°], from which ten specimens were cut in two directions (5 transverse (TR) and 5 longitudinal (LG)), Figure 3.5.

The samples fabricated and pre-stressed by cyclic fatigue stress after a pulsating cycle can be seen in Figure 3.6.

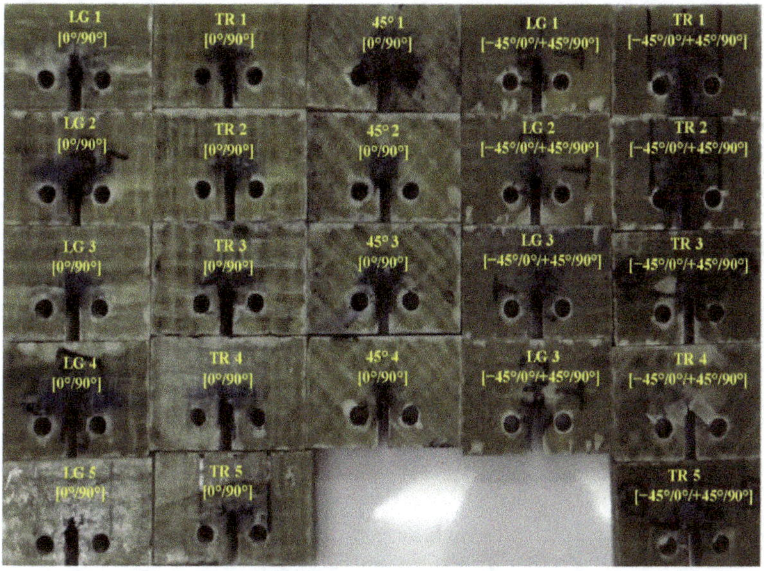

Fig. 3.6 Samples from GFRP plates

3.4 Testing for the determination of da/dN

After sampling as described above, the GFRP samples were subjected to tensile testing to determine the rate of crack propagation. Figure 3.7 shows how the GFRP CT specimen was subjected to tensile testing using the Instron 8801 testing machine.

Fig. 3.7 Tensile testing using the Instron 8801

Additionally, on the same machine, but using the WaveMatrix software, pulsed tensile fatigue stresses were performed to pre-shape the CT specimens with lateral notch, Figure 3.8. The maximum force at pre-shape was 5600 N corresponding to a stress (not taking into account the stress concentration factor) of about 35 MPa.

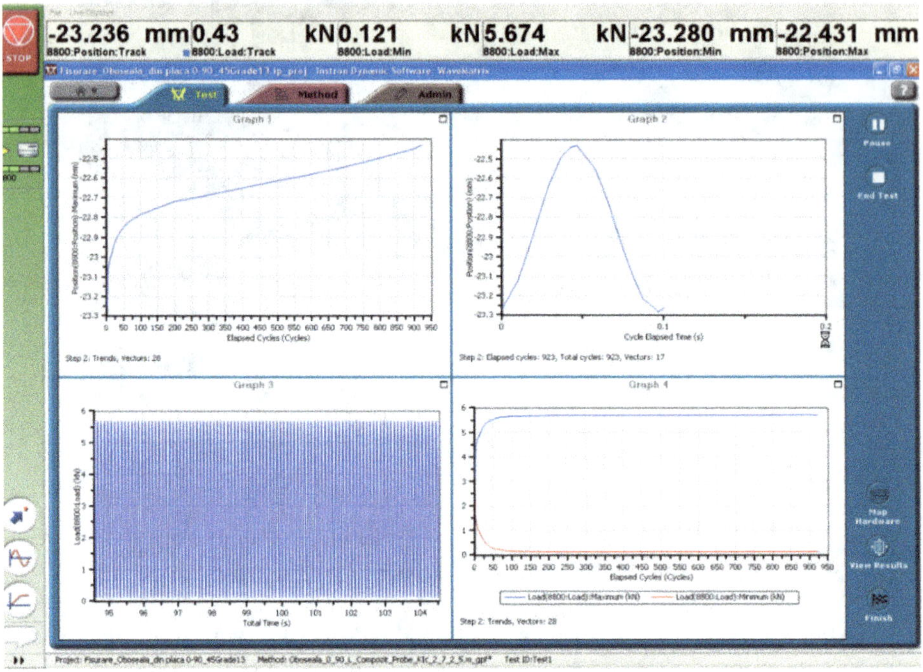

Fig. 3.8 GFRP pre-testing program

In the framework of the "da/dN" software, if the calculation method is chosen using the Compliance, from the main menu you can check the coefficients to be used in relation (3.4) for the calculation of the propagated crack length a. The compliance in relation (3.2) uses the values given in the menu in Figure 3.9 for E, B and B_{net}. The value for v (crack flank opening displacement) is taken automatically from the strain gauge mounted on the front face of the specimen, see Figure 3.7, (COD - crack opening displacement) and the force is taken from the program run which also appears in the secondary menu - when the test takes place.

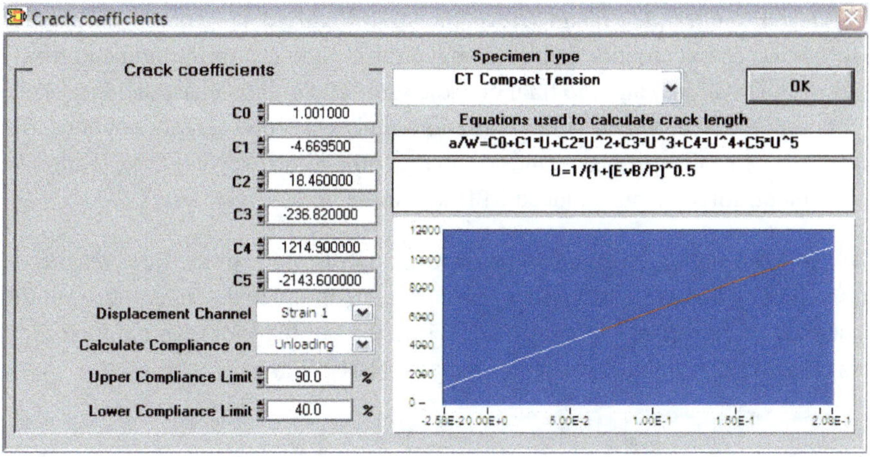

Fig. 3.9 Coefficients used for the Compliance method

Figure 3.10 shows the working parameters to be provided by the "da/dN" software in relation to the geometry of the CT specimen and the elastic and mechanical characteristics of the material.

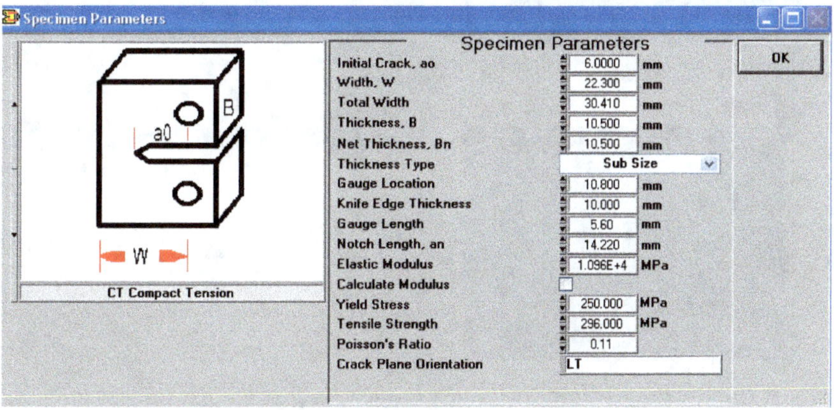

Fig. 3.10 Working parameters - geometry and material characteristics

It should be noted that these characteristics have been previously determined for each direction and the values thus obtained have been entered here, in the same way, in relation to each plate from which the sample was taken and corresponding to each cutting direction.

During the test, the menu screen shown in Figure 3.11 appears. On this screen, the following characteristics can be changed during the test: minimum and maximum values of the test force and the final crack length. The graphs that can appear here are: the graph of force versus time variation and a second graph contains several possibilities for the abscissa and ordinate: unloading versus time, force versus displacement, force versus reduced LPD and force versus hysteresis LPD.

As illustrated in the figure below, the maximum value of the force reaches 16662 N at the point when 26638 stress cycles have been completed. Under these conditions, at that point of time, the stress introduced into the sample (not taking into account the stress concentration factor) was approximately 104 MPa.

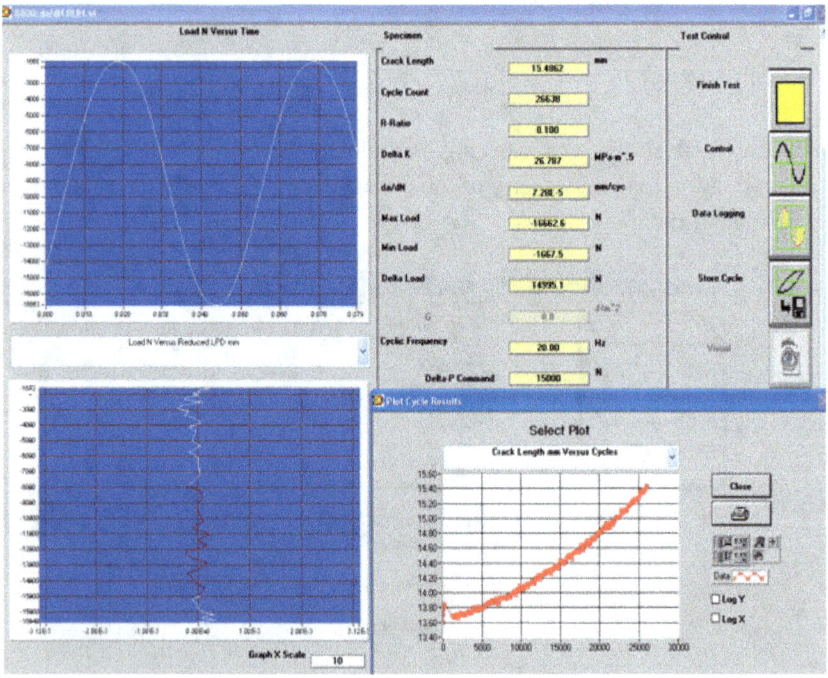

Fig. 3.11 Test screen - secondary

The results are presented both as graphs and data tables, and are as follows:

- Variation of crack length versus number of cycles;
- Variation of da/dN versus ΔK;
- Variation of crack flank displacement versus number of cycles;
- Variation of da/dN versus number of cycles;

- Variation of ΔK versus number of cycles;
- Data table.

3.5 Experimental results

In the following, the experimental results obtained when testing the GFRP-made compact specimens to determine the fatigue crack growth rate are presented.

From the data table provided by the testing machine, Table 3.1, the necessary data is taken and some of the graphs presented above are plotted.

Table 3.1 Table of data supplied by the testing machine

Cycles	Crack Length	EvBP	da/dN mm/cyc	Delta-P N	Delta-K MPa-m	R-ratio	Active Pd	Ref Pd Vo	G J/m^2	Date/Time
1.00E+00	3.18E+00	1.48E+01	0.00E+00	4.90E+02	0.00E+00	7.13E-01	0.00E+00	0.00E+00	0.00E+00	20230215/15:20:58
1.10E+01	3.36E+00	1.51E+01	0.00E+00	4.24E+02	0.00E+00	7.53E-01	0.00E+00	0.00E+00	0.00E+00	20230215/15:20:59
2.20E+01	3.65E+00	1.57E+01	0.00E+00	4.87E+02	0.00E+00	7.40E-01	0.00E+00	0.00E+00	0.00E+00	20230215/15:21:00
7.70E+01	3.67E+00	1.58E+01	6.82E-04	7.95E+02	1.96E+00	6.71E-01	0.00E+00	0.00E+00	0.00E+00	20230215/15:21:06
8.80E+01	3.68E+00	1.58E+01	1.45E-03	8.46E+02	2.07E+00	6.63E-01	0.00E+00	0.00E+00	0.00E+00	20230215/15:21:07
9.80E+01	3.68E+00	1.58E+01	7.53E-04	9.00E+02	2.20E+00	6.51E-01	0.00E+00	0.00E+00	0.00E+00	20230215/15:21:08
1.19E+02	3.69E+00	1.58E+01	9.15E-04	1.01E+03	2.49E+00	6.29E-01	0.00E+00	0.00E+00	0.00E+00	20230215/15:21:10
1.40E+02	3.75E+00	1.60E+01	9.07E-04	1.13E+03	2.77E+00	6.09E-01	0.00E+00	0.00E+00	0.00E+00	20230215/15:21:12
1.71E+02	3.75E+00	1.60E+01	1.14E-03	1.29E+03	3.18E+00	5.78E-01	0.00E+00	0.00E+00	0.00E+00	20230215/15:21:15
1.81E+02	3.76E+00	1.60E+01	1.22E-03	1.34E+03	3.31E+00	5.70E-01	0.00E+00	0.00E+00	0.00E+00	20230215/15:21:16
1.92E+02	3.76E+00	1.60E+01	1.01E-03	1.40E+03	3.46E+00	5.60E-01	0.00E+00	0.00E+00	0.00E+00	20230215/15:21:17
2.02E+02	3.81E+00	1.61E+01	1.54E-03	1.44E+03	3.59E+00	5.50E-01	0.00E+00	0.00E+00	0.00E+00	20230215/15:21:18
2.53E+02	3.87E+00	1.62E+01	1.10E-03	1.69E+03	4.23E+00	5.08E-01	0.00E+00	0.00E+00	0.00E+00	20230215/15:21:23
2.95E+02	3.89E+00	1.63E+01	7.91E-04	1.87E+03	4.73E+00	4.75E-01	0.00E+00	0.00E+00	0.00E+00	20230215/15:21:27
3.26E+02	3.92E+00	1.64E+01	4.83E-04	2.01E+03	5.08E+00	4.52E-01	0.00E+00	0.00E+00	0.00E+00	20230215/15:21:31
4.70E+02	3.93E+00	1.64E+01	2.28E-04	2.55E+03	6.46E+00	3.64E-01	0.00E+00	0.00E+00	0.00E+00	20230215/15:21:45
5.00E+02	3.94E+00	1.64E+01	3.01E-04	2.64E+03	6.70E+00	3.50E-01	0.00E+00	0.00E+00	0.00E+00	20230215/15:21:48
5.83E+02	3.95E+00	1.64E+01	3.08E-04	2.88E+03	7.34E+00	3.11E-01	0.00E+00	0.00E+00	0.00E+00	20230215/15:21:56
6.04E+02	3.98E+00	1.65E+01	3.59E-04	2.93E+03	7.48E+00	3.05E-01	0.00E+00	0.00E+00	0.00E+00	20230215/15:21:58
6.46E+02	4.00E+00	1.65E+01	3.62E-04	3.03E+03	7.75E+00	2.89E-01	0.00E+00	0.00E+00	0.00E+00	20230215/15:22:03

The graphs were drawn for five cases (types of specimens) considered and visible in Figure 3.6. These specimens, in relation to the reinforcement mode and the cutting mode of the specimens from the two composite plates, were named as follows:

- [-45°/0°/+45°/90°] - longitudinal (cutting direction);
- [-45°/0°/+45°/90°] – transversal (cutting direction);
- [0°/90°] – longitudinal (cutting direction);
- [0°/90°] – transversal (cutting direction);
- [0°/90°] – diagonal (cutting direction);

Of the four or five samples of the same type available, graphs were drawn only for the one that was more representative in terms of the results obtained. The graphs drawn are:

- variation of crack length versus number of cycles, a-N;
- the variation of fatigue crack growth rate versus number of cycles, da/dN-N;
- the variation of the fatigue crack growth rate versus the variation of the stress intensity rage, da/dN-ΔK.

Specimen LG 4 [-45°/0°/+45°/90°] – longitudinal cut

Figures 3.12a and 3.12b show the variation of crack length as a function of the number of cycles, for the specimen with the fiber orientation at [-45°/0°/+45°/90°] cut longitudinally. The longitudinal displacement occurred along the geometric axis of the sample, which was aligned parallel to the orientation of the fibers constituting the warp of the fabric. We have two representations, respectively the one taken from the testing machine and an Excel representation in which the initial data at the start of the test and the final data at the stop of the test have been removed. There is an initial fatigue propagated crack length of 3.65 mm for this sample. It is observed that the crack length increases sharply at the beginning, after which its variation with respect to the number of cycles stabilizes after an approximately constant slope. Since we are interested in the crack propagation velocity in the zone of stable propagation, about 10000 stress cycles were sufficient. After a relatively large number of stress cycles (>>10000), the crack propagation will occur with accelerated speed. We are not interested in this particular aspect, because it is important for the components working in service not to reach this stage.

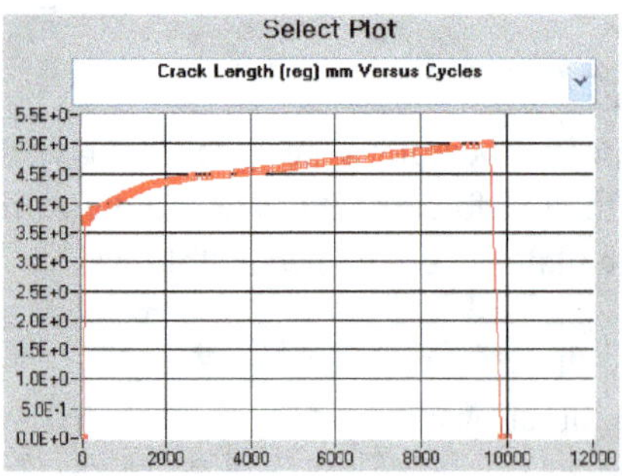

a) image taken from the PC of the testing machine

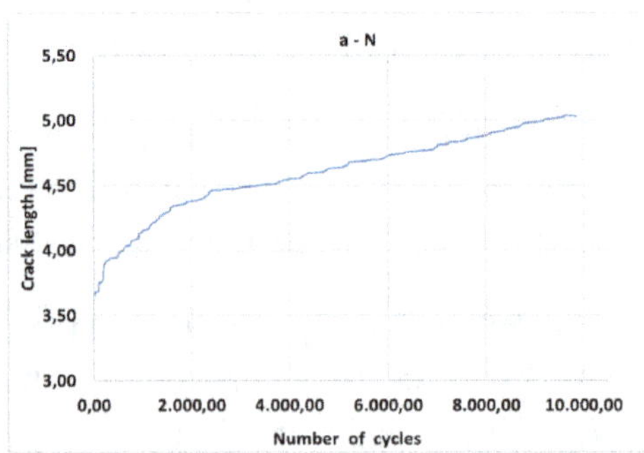

b) a-N variation, processed in Excel

***Fig. 3.12** Variation of crack length versus number of cycles for the [-45°/0°/+45°/90°] sample longitudinal*

Figure 3.13 shows the variation of the fatigue crack growth rate da/dN, in relation to the number of cycles. The fatigue crack growth rate varies during the test. For example, during the final period of the test the fatigue crack growth rate increases significantly. However, we are interested in the area where the fatigue crack growth rate has a constant value, when we have the crack length at an intermediate value: neither at the beginning when phenomena related to hardening appear, nor at the end when the crack length has already reached a value at which it propagates with a fairly high speed, even suddenly. Consequently, we can consider for the fatigue crack growth rate, calculated with the relation in figure 3.13b at 4500 cycles, the value:

$$da/dN = 0.000103 \text{ mm/cycle}$$

as being representative of this sample.

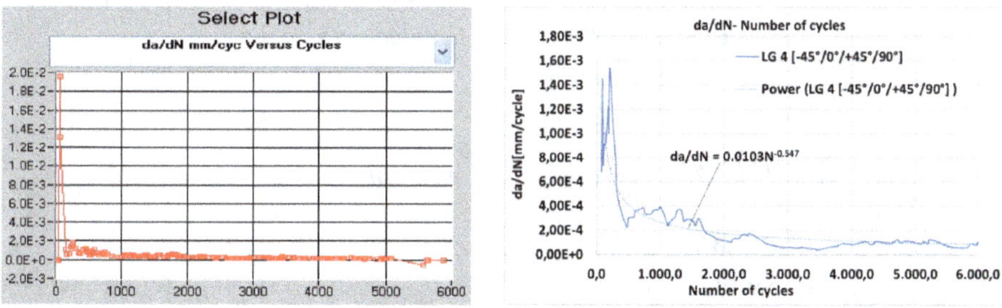

a) image taken from the PC of the testing machine

b) da/dN-a variation, processed in Excel

Fig. 3.13 Variation of fatigue crack growth rate versus number of cycles for the [-45°/0°/+45°/90°] longitudinal sample

From table 3.1, the data can be taken and the diagram of the fatigue crack growth rate in relation to the variation of the stress intensity rage, ΔK, can be drawn, Figure 3.14. And in this case, for Figure 3.14b, with the variant plotted in Excel, the first and last points were not taken into account. For the initial part of the load, an increase and then a decrease of this variation are observed. After a certain number of cycles, the variation of da/dN with respect to ΔK stabilizes at the value specified above. It is evident that ΔK increases continuously as the length of the propagated crack increases, as shown in Figure 3.12.

For the previous value of the crack propagation velocity, using the logarithmic relationship in figure3.14b, $\left[ln\left(\frac{da/dN}{0.007}\right)\right] \cdot \left(\frac{1}{-1.651}\right) = ln\,(\Delta K)$, the variation of the stress intensity rage is calculated as:

$$\Delta K = 12{,}84\ MPa \cdot \sqrt{m}$$

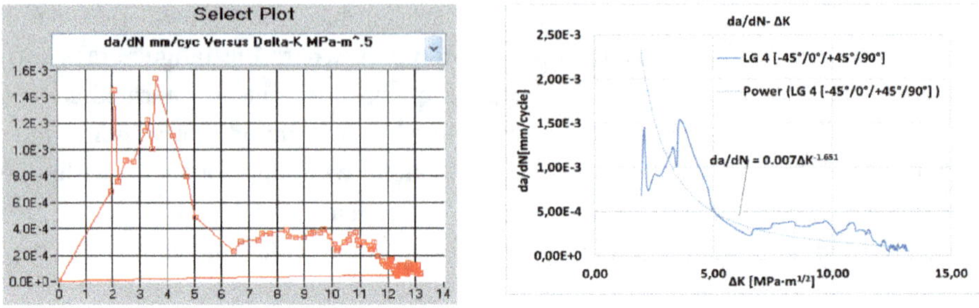

a) **image taken from the PC of the testing machine**

b) **da/dN-ΔK a variation, processed in Excel**

Fig. 3.14 Variation of fatigue crack growth rate versus the variation of the stress intensity rage for the [-45°/0°/+45°/90°] longitudinal sample

Specimen TR 3 [-45°/0°/+45°/90°] – transversal cut

Figure 3.15 shows the variation of the crack length in relation to the number of cycles for the sample with the fiber arrangement [-45°/0°/+45°/90°] – cut transversely (with the sample axis parallel to the weft).

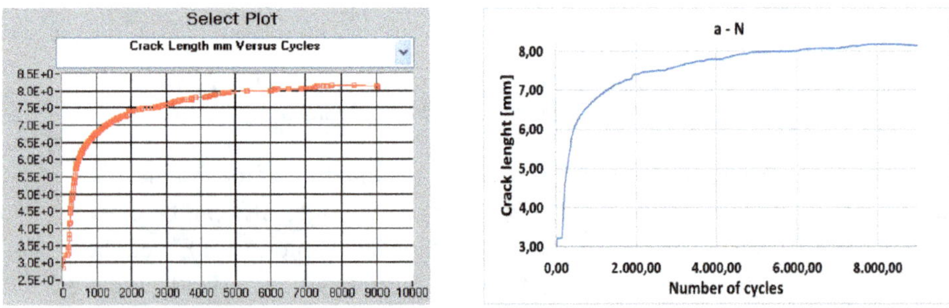

a) **image taken from the PC of the testing machine**

b) **a-N variation, processed in Excel**

Fig. 3.15 Variation of crack length versus number of cycles for the [-45°/0°/+45°/90°] transversal sample

An even more pronounced increase in crack length is observed in a shorter time (highlighted by the number of loading cycles). After only 1000 cycles, the crack grows from approximately 3 mm to 6.7 mm. After 2000 cycles, when the crack has reached approximately 7.3 mm, a slower crack growth occurs with respect to the number of cycles. As a result, this is our area of interest that we will identify in the variation of the crack propagation velocity in Figure 3.16.

Figure 3.16 shows the variation of fatigue crack growth rate, da/dN, in relation to the number of cycles. Figure 3.15a illustrates that after a larger number of cycles, the fatigue crack growth rate stabilizes at a certain value, with small variations around it. As a result, we can calculate the fatigue crack growth rate with the relationship presented in Figure 3.16b, for N=5000 cycles. The result will be:

da/dN= 0.000122 mm/cycle

value considered representative for this sample.

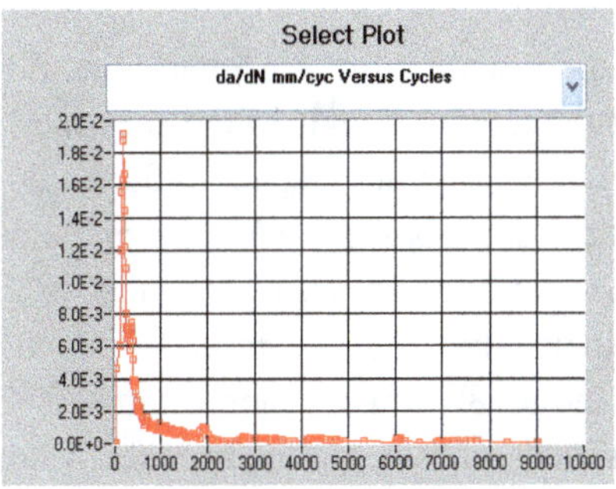

a) image taken from the PC of the testing machine

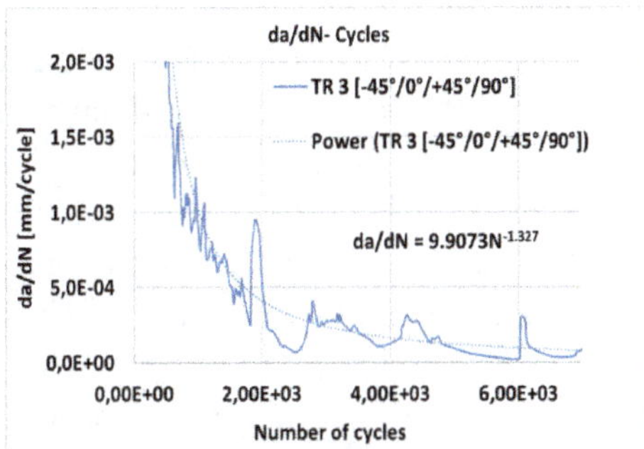

b) da/dN-a variation, processed in Excel

Fig. 3.16 Variation of fatigue crack growth rate versus number of cycles for the [-45°/0°/+45°/90°] transversal sample

In Figures 3.17a and 3.17b, the variation of the crack propagation speed is observed in relation to the variation of the stress intensity factor. It is found that, after a certain time, this variation stabilizes. In any case, we will calculate the variation of the stress intensity factor in relation to the crack propagation speed previously determined. For this purpose, we have the calculation relationship provided in Figure 3.17b, proceeding in the same way as in the previous test. Thus, we will have:

$$\Delta K = 22{,}28\, MPa \cdot \sqrt{m}$$

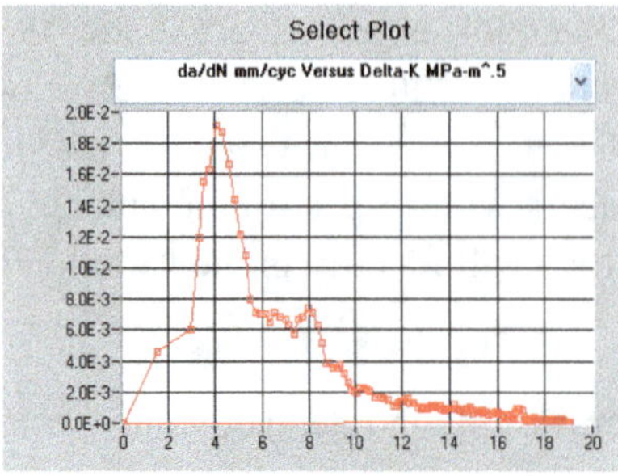

a) **image taken from the PC of the testing machine**

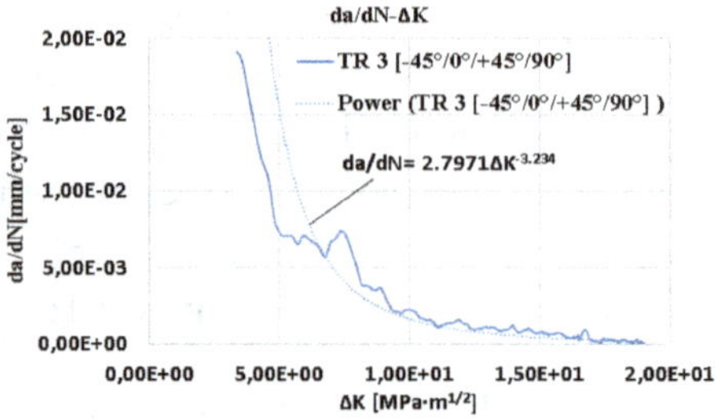

b) **da/dN-ΔK variation, processed in Excel**

Fig. 3.17 Variation of fatigue crack growth rate versus the variation of the stress intensity rage for the [-45°/0°/+45°/90°] transversal sample

Specimen LG 1 [0°/90°]– longitudinal cut

In Figures 3.18a and 3.18b, for the specimen taken from the plate with fiber orientation at [0°/90°] with longitudinal cutting (the geometric axis of the specimen is parallel to the warp), we have the variation of crack length with respect to the number of cycles, in the two graphical variants. In these figures, we observe a more pronounced slope of crack length increase with respect to the number of cycles. Basically, in the interval of only 2200 cycles, the crack propagated by about 1 mm. With small deviations, the variation shown in Figures 3.18a and 3.18b, remains approximately the same. Under these conditions, it can be said that this area is representative for further calculation of the fatigue crack growth rate for the [0°/90°]- sample.

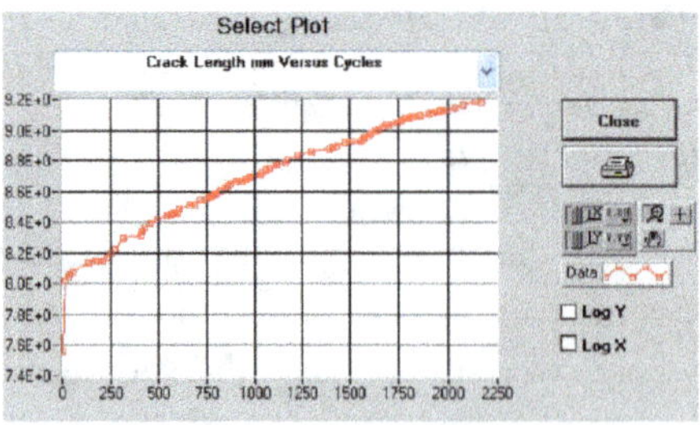

a) **image taken from the PC of the testing machine**

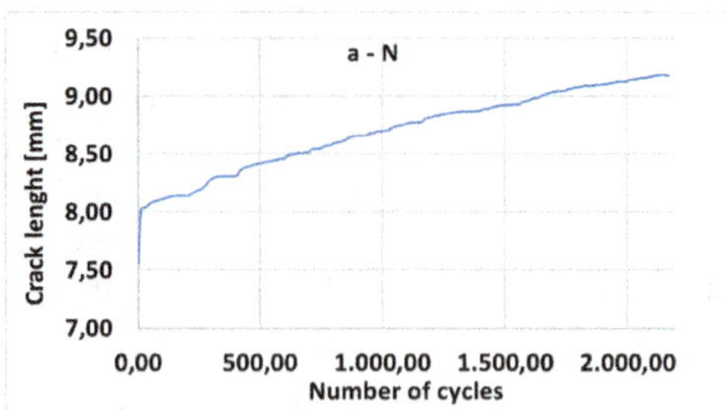

b) **a-N variation, processed in Excel**

Fig. 3.18 Variation of crack length versus number of cycles for the [0°/90°] longitudinal sample

Figure 3.19 shows the variation of fatigue crack growth rate, da/dN, versus the number of cycles. In Figure 3.19b, the stress onset points have been removed in order to be able to plot the trend curve as close as possible to the variation in the area of interest, i.e. the area of stable crack propagation. In Figure 3.19a it can be seen that, after a larger number of cycles, the fatigue crack growth rate has a slight decreasing trend with small variations. From the observation of this variation, we can consider as a representative value for the fatigue crack growth rate for the [0°/90°] longitudinal [0°/90°] specimen the one which is recorded at the value of N=1000 cycles. As a result, the fatigue crack growth rate can be calculated with the relation shown in Figure 3.19b. It will therefore result:

da/dN= 0.000522 mm/cycle

value considered to be representative for the [0°/90°] longitudinal sample.

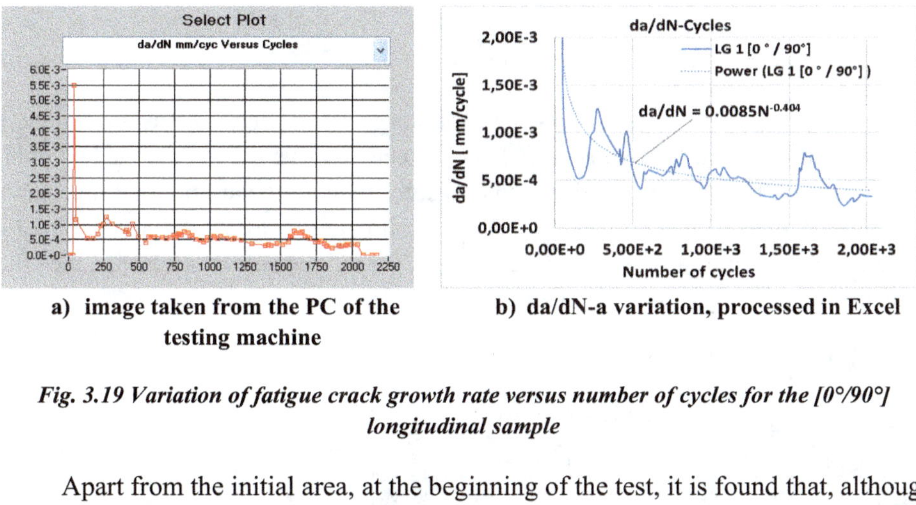

a) image taken from the PC of the testing machine **b) da/dN-a variation, processed in Excel**

Fig. 3.19 Variation of fatigue crack growth rate versus number of cycles for the [0°/90°] longitudinal sample

Apart from the initial area, at the beginning of the test, it is found that, although ΔK increases (as the propagated crack length increases), the fatigue crack growth rate shows slight variations without significant jumps. The variation of the stress intensity rage is calculated with respect to the magnitude of the fatigue crack growth rate determined previously. In order to achieve this, we have at our disposal the calculation relation provided in Figure 3.20b, proceeding in the same way as for the previous sample. Thus, we have:

$$\Delta K = 14{,}74\, MPa \cdot \sqrt{m}$$

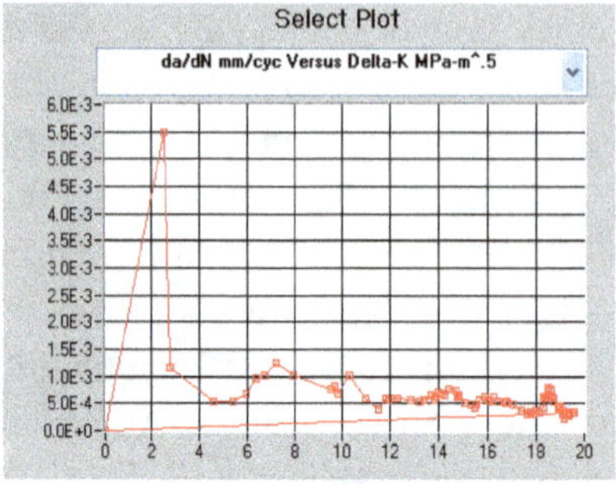

a) **image taken from the PC of the testing machine**

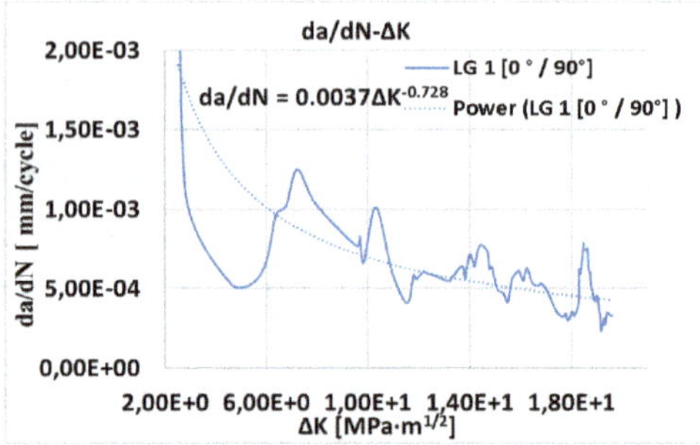

b) **da/dN-ΔK variation, processed in Excel**

Fig. 3.20 Variation of fatigue crack growth rate versus the variation of the stress intensity rage for the [0°/90°] longitudinal sample

Specimen TR 1 [0°/90°] – transversal cut

Figures 3.21a and 3.21b show the variation of crack length versus the number of cycles for the specimen oriented at [0°/90°]. It can be observed that, at the beginning, the crack length increases relatively steeply after which the variation with respect to the number of cycles stabilizes after an approximately constant slope. The area of interest being the variation with constant slope, we will be able to determine the crack propagation velocity representative for this sample.

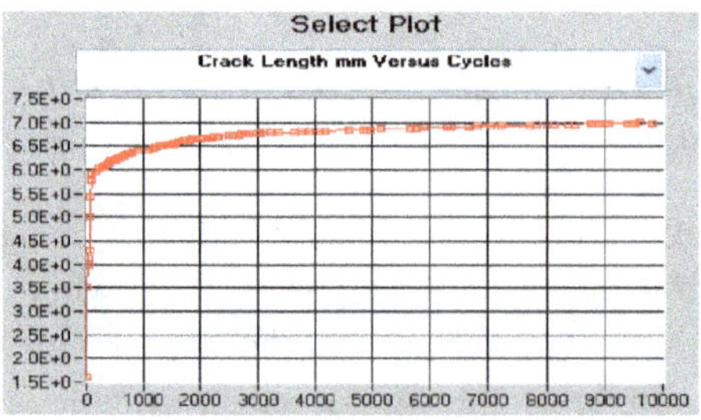

a) **image taken from the PC of the testing machine**

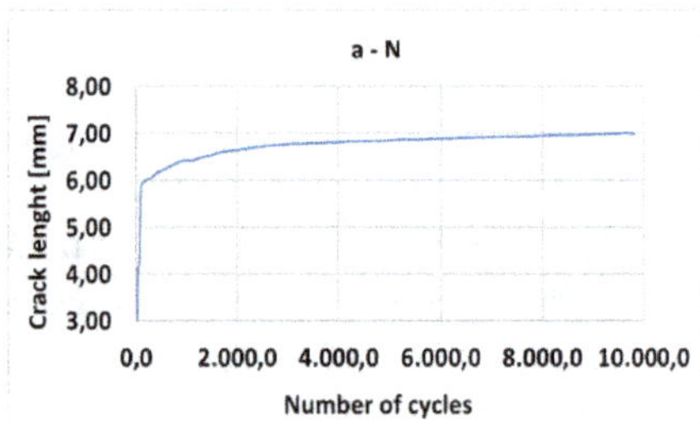

b) **a -N variation, processed in Excel**

Fig. 3.21 Variation of crack length versus number of cycles for the [0°/90°] transversal sample

Figure 3.22 plots the variation of the fatigue crack growth rate, da/dN, versus the number of cycles. Considering also the variation of crack length with approximately constant slope in Figure 3.21b, from Figure 3.22a it can be seen that after a larger number of stress cycles the fatigue crack growth rate stabilizes. It is obvious that, if we kept the stress for a much larger number of cycles, the fatigue crack growth rate would increase, but we would be in the area of sudden crack propagation, which is not of interest, as our purpose is not to get into this situation. Consequently, we can consider for the crack propagation velocity, calculated at N=4000, the value

$$da/dN = 5.828E\text{-}05 \text{ mm/cycle}$$

as being representative of the sample taken from the [0°/90°] cross-cut plate.

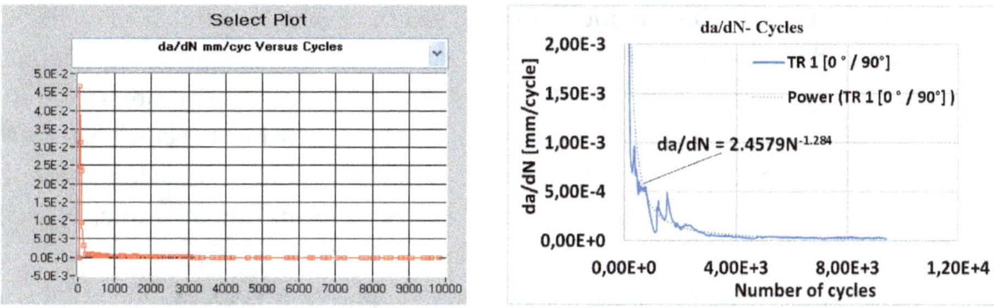

a) image taken from the PC of the testing machine **b) da/dN-a variation, processed in Excel**

Fig. 3.22 Variation of fatigue crack growth rate versus number of cycles for the [0°/90°] transversal sample

Next, the variation of the fatigue crack growth rate da/dN is considered in relation to the stress intensity rage ΔK. Figure 3.23a shows this variation as taken from the testing machine. Figure 3.23b plots the variation without the test starting points in order to establish an appropriate variation relationship between the two quantities.

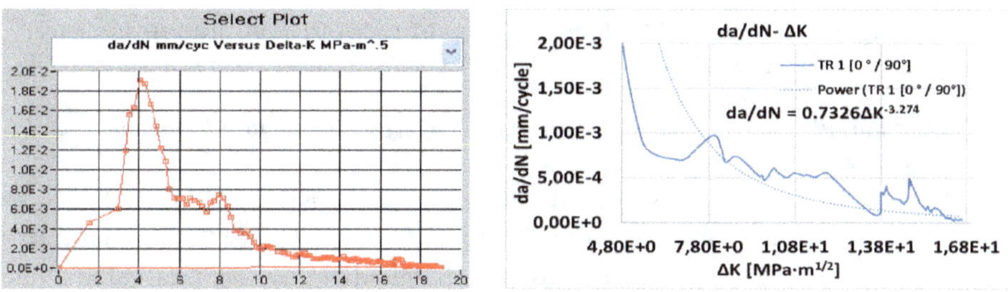

a) image taken from the PC of the testing machine **b) da/dN-ΔK variation, processed in Excel**

Fig. 3.23 Variation of fatigue crack growth rate versus the variation of the stress intensity rage for the [0°/90°] transversal sample

Using the relation given in Figure 3.23b, and the value for da/dN established above, for the variation of the stress intensity factor after N=4000 stress cycles, we have:

$$\Delta K = 17{,}71 MPa \cdot \sqrt{m}$$

Specimen 4 [0°/90°] – diagonal cut

Figure 3.24 shows the variation of crack length versus the number of cycles for the specimen with fiber orientation at [0°/90°] with the geometric axis of the specimen, creating a 45° angle with the warp/weft. In this case as well a more pronounced initial increase in crack length is observed. After about 2000 cycles, when the crack has grown from 3.18 mm to about 4.36 mm, the slope of the curve (tangent at each point to the variation curve) remains approximately constant. As a result, the area after N=2000 cycles is of interest to us and we will consider it further in order to calculate the representative value of this sample for the crack propagation velocity, identifiable in Figure 3.25.

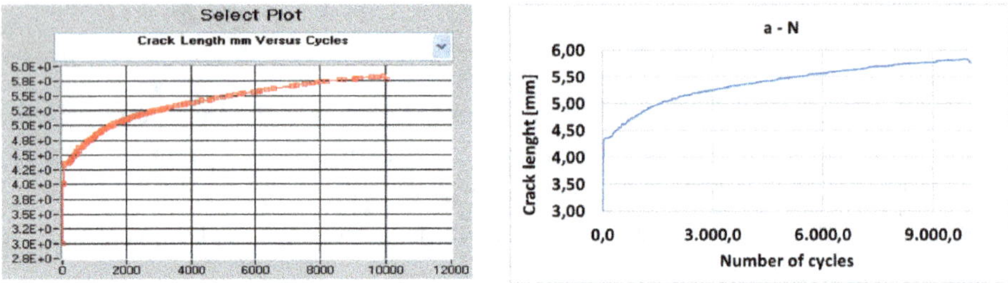

a) image taken from the PC of the testing machine **b) a -N variation, processed in Excel**

Fig. 3.24 Variation of crack length versus number of cycles for the [0°/90 diagonal sample]

Figures 3.25a and b show the variation of fatigue crack growth rate, da/dN, with respect to the number of cycles. From Figure 3.25a it can be seen that, after a larger number of cycles, the fatigue crack growth rate stabilizes at a certain value without much variation around it. As a result, the fatigue crack growth rate can be calculated with the relation shown in Figure 3.25b, for N = 5000 cycles. It will result:

da/dN= 0.0001169 mm/cycle

value considered to be representative of the [0°/90°]- diagonal sample.

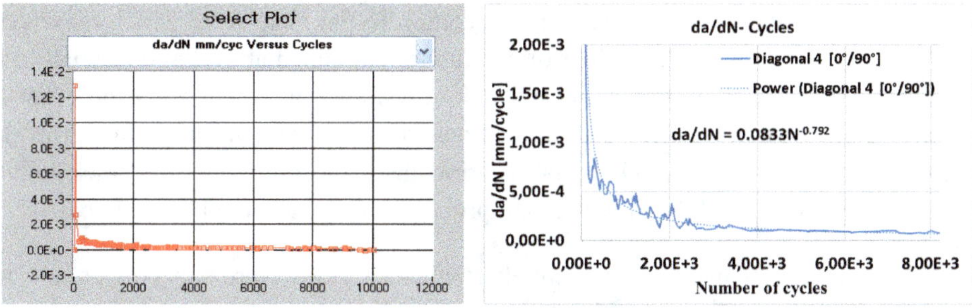

a) image taken from the PC of the testing machine

b) da/dN-a variation, processed in Excel

Fig. 3.25 Variation of fatigue crack growth rate versus number of cycles for the [0°/90°] diagonal sample

Besides the initial area, at the beginning of the test, it is found that, although ΔK increases (as the propagated crack length increases), the fatigue crack growth rate shows slight variations without significant jumps. The variation of the stress intensity rage is calculated with respect to the magnitude of the fatigue crack growth rate determined previously. For this, we have the calculation relation given in figure 3.26b. Thus, we have:

$$\Delta K = 14{,}90\ MPa \cdot \sqrt{m}$$

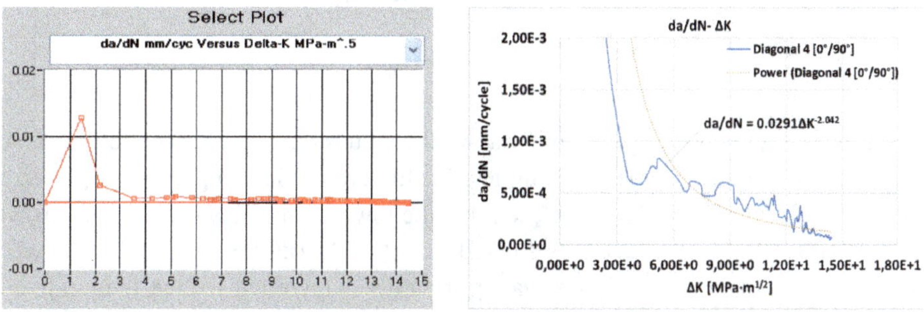

a) image taken from the PC of the testing machine

b) da/dN-ΔK variation, processed in Excel

Fig. 3.26 Variation of fatigue crack growth rate versus the variation of the stress intensity rage for the [0°/90°] diagonal sample

For the five cases of specimen types presented above, the fatigue crack growth rate, da/dN [mm/cycle], and the variation of the stress intensity rage at the time when da/dN was calculated as representative were determined.

The samples were differentiated in relation to the plate from which they were cut, the plates having a certain fiber orientation as explained above, and in relation to the direction of sampling from the plate. Five samples were cut for each case. In general, for all samples there is a more accelerated increase in crack length at the beginning of the test followed by a stabilization, the relationship between crack length, **a**, and number of cycles, **N**, in the second propagation stage being almost linear.

The number of cycles was relatively small because we were only concerned with the zone of stable crack propagation that we obtained. Table 3.2 summarizes the values obtained for the fatigue crack growth rate, da/dN, and, correspondingly, for the variation of the stress intensity rage.

Tabel 3.2 Values obtained for da/dN and ΔK

Nr. crt.	Specimen name	da/dN [mm/cycle]	ΔK [$MPa \cdot m^{1/2}$]
1	Plate [-45°/0°/+45°/90°] – specimen cut longitudinal LG4	0,000103	12,84
2	Plate [-45°/0°/+45°/90°] – specimen cut transversal TR3	0,000122	22,28
3	Plate [0°/90°] – specimen cut longitudinal LG1	0,000522	14,74
4	Plate [0°/90°] – specimen cut transversal TR1	$8{,}43 \cdot 10^{-5}$	15,96
5	Plate [0°/90°] – specimen cut diagonal	0,0001169	14,90

Figure 3.27 shows a graph illustrating the values obtained for the fatigue crack growth rate for the five types of samples. In this graph the samples were placed in ascending order of da/dN. It can be seen that the lowest value of fatigue crack growth rate is recorded for the transversely flowed sample TR3 (in the direction of the weft) from plate [0°/90°] and the highest value of da/dN is recorded for the longitudinally flowed sample LG4 (in the direction of the warp) from plate [0°/90°]. In any case, the values determined for the fatigue crack growth rate for the five cases are close to each other, this aspect proving the correctness of the determinations.

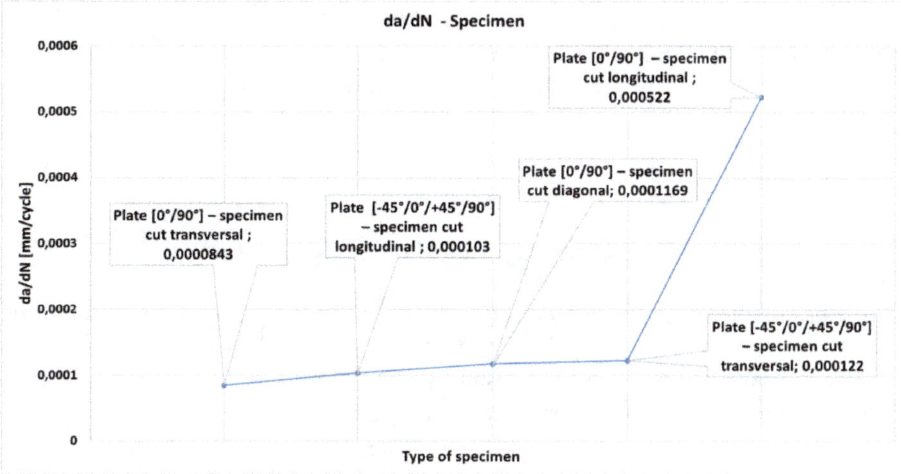

Fig. 3.27 Values obtained for da/dN

Figure 3.28 shows the values for the variation of the stress intensity rage. They have been put in ascending order just for a simpler evaluation of the differences that occurred.

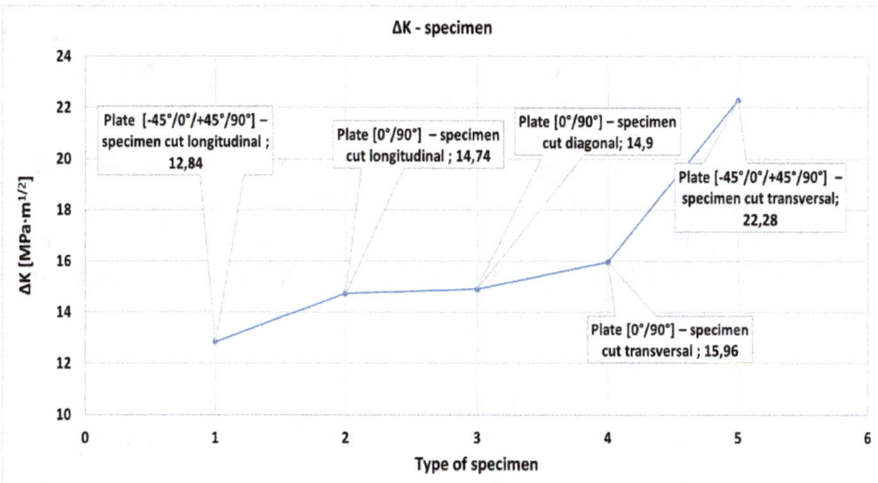

Fig. 3.28 Values obtained for ΔK

Here, however, there is a significant difference between the longitudinally flowed sample TR3 from plate [-45°/0°/+45°/90°] (12.84 MP-m$^{1/2}$) compared to the transversely flowed sample TR1 from plate [-45°/0°/+45°/90°] (22.28 MP-m$^{1/2}$).

However, the values for the variation of the stress intensity rage, being calculated pointwise, when the fatigue crack growth rate stabilizes and even at the point at which it is calculated, are not entirely relevant in terms of the differences from one specimen to another.

3.6 Defectoscopic determinations on CT samples of GFRP

3.6.1 Method used to determine the initial crack propagation through fatigue

In this subchapter, liquid penetrant defectoscopy applied on Compact Tension specimens made of GFRP is presented. The configuration and dimensions of the specimens were presented earlier in this chapter. They were fabricated in order to determine some Fracture Mechanics characteristics, namely, fatigue crack growth rate and crack toughness. The cutting of the specimens from the plates was carried out in two or three different directions, depending on the type of reinforcement of the plate, in order to highlight the differences in the subsequently determined sizes.

In these calculations, the initial fatigue propagated crack length is introduced in the initial data provided to the dedicated software that determines the propagated crack length which is also related to the initial crack length. Consequently, the initial fatigue propagated crack length is a very important parameter for the correctness of the propagated crack length calculation in the actual tests.

During and after the fatigue pre-curing it was found that for some samples, taken in different directions from specific GFRP plates, it was visually or optically quite difficult to determine the initial crack length. Under these conditions, the only possibility to determine the fatigue stress-initiated crack length was with penetrating liquids.

3.6.2 Determination of cracks in CT samples using penetrant liquids

3.6.2.1 Peculiarities of the penetrant inspection method

The liquid penetrant control method has been known and applied in a rudimentary form (e.g. petroleum with calcium carbonate powder mixed with alcohol) for over a century. Modern techniques, in a similar manner to those used today, have been known since around the Second World War. Further refinements have their origins in the development of aviation, rocketry, nuclear and aerospace technology.

Liquid penetrant inspection highlights any surface discontinuities (imperfections). It can be applied to any material, shape and size of part, in shop or site conditions, on dry surfaces, at temperatures above 10 ÷ 15°C.

The method is productive, inexpensive, easy to use and also suitable for inspection of large areas (lengths). The results are conclusive, immediate and easy to interpret. Indications from discontinuities are increased by absorbing the penetrant several times.

Control with penetrating liquids involves chemical cleaning of the surface of impurities, post-operative washing operations, especially when applied interfaced, during the successive deposition of weld layers. Limitations arise in some cases due to the fact that some liquid penetrant recipes utilize flammable and toxic materials, which require appropriate washing and evacuation measures. The finer the surface finish of the area under examination, the higher the probability of detecting defects. The control method is relevant for discontinuities in the range 0.1 - 5 mm. At peak performance, the level of detectability goes down to below one micron. Liquid penetrant inspection should be seen as a refinement and extension of visual examination. Liquid penetrant inspection of welded joints involves, like magnetic powder inspection, the examination of areas of at least 20 to 30 mm on either side along the area of interest, which has the advantage of also revealing any cracks propagating in the adjacent material. It can be applied at different stages of execution. Limitations in the application of the method can only be caused by temperature, as ordinary penetrants cannot be used at temperatures exceeding 50°C. There are also special liquid penetrants with flash points above 250°C, which makes it possible to check between layers when welding or plating.

For low temperature control, penetrants have been developed that can be applied down to - 35°C. Liquid penetrant control is also used with good results for plating and metallizing. The area of application is practically not limited by the type of material and can be used on all types of steels, cast iron, aluminum and magnesium alloys and, in general, non-ferrous metals, as well as amorphous materials, plastics, ceramics, glass, etc.

3.6.2.2. Principle and methods of control

Penetrating liquid inspection consists of applying a liquid with good penetration qualities into surface discontinuities on the surface to be inspected and highlighting them by contrast with a developer, Figure 3.29. Penetration into the finest discontinuities - pores, cracks, etc. - occurs due to capillary action. Penetrant development occurs due to the absorption effect of the developer.

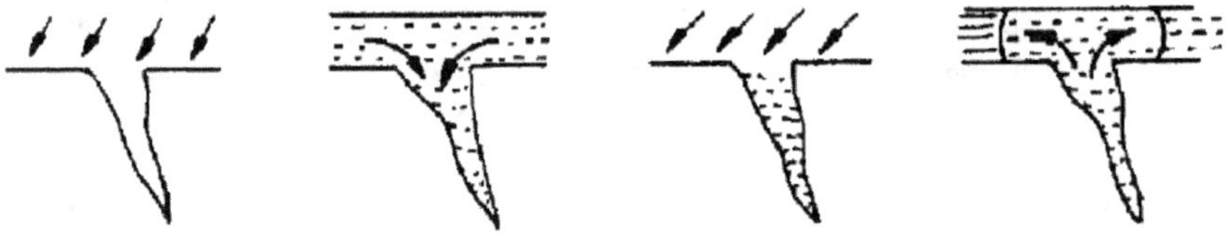

a b c d

Fig. 3.29 The principle of liquid penetrant control: a. - surface cleaning: b. - penetrant application and infiltration in discontinuities: c. - removal of excess penetrant: d - application of developer and absorption of penetrant (Gh. Amza et.al., 2011)

Liquid penetrant inspection exclusively highlights open surface discontinuities, such as pores, cracks, overlaps, lack of penetration, narrow open surface, marginal notches, exfoliations in the base material, craters. Especially relevant for the inspection with penetrating liquids are single pores and cracks, either thermal or fatigue cracks, which in most cases are difficult to detect by visual inspection. Porous surfaces, or areas with high pore density or very rough areas, cannot be effectively inspected because of the difficulty in interpreting the revealed indications. In principle, parts should be inspected before applying heat treatments or chipping as the latter, in particular, can mask or even close finer discontinuities.

The main methods of control with penetrant liquids are as follows:

a) The coloring method, in which the contrast for revealing the discontinuities is one of color, usually red on a white background, is the most frequently used;
b) The fluorescent method in which the contrast for revealing discontinuities is obtained by the glow observed in ultraviolet light and black examination field; the contrast is usually yellow-green on a dark purple background;
c) The ultrasonic activation method, where the excess infiltration energy of the penetrant used is provided by ultrasonic vibrations and emission of elastic waves into the penetrating or post-emulsion medium. A lower frequency

increases the penetration capability, but it may have disruptive effects below 20 kHz. Ultrasonic activation is mainly applied in the case of product tightness control by helping to overcome the adhesion force and forcing the penetration process. With good results, it is used to wash capillary surfaces of contaminated materials. Vibration also helps to minimize the penetration time.

Regardless of the type of penetrant or developer, liquid penetrant testing involves the following operations:

1) surface preparation;
2) penetrant application;
3) removal of excess penetrant;
4) applying the developer;
5) surface examination and interpretation of results;
6) marking of defective areas on the part.

In the case of using penetrants with post-emulsification, an additional phase of adding the emulsifying agent is required, after the penetration time of the penetrant has expired.

3.6.2.3 Penetrant development mechanism

The penetrant remaining in the discontinuities, after removing the excess from the surface of the examined part, must be absorbed on the surface to obtain, through color contrast or fluorescent contrast, the indication of the defect. This role is fulfilled by the developer, which provides the following functions, as it can be observed in Figure 3.30:

1. absorption of the remaining penetrant trapped in discontinuities;
2. fixing the penetrant and widening the indication as the layer thickness increases;
3. forming and improving the contrast between penetrant and developer.

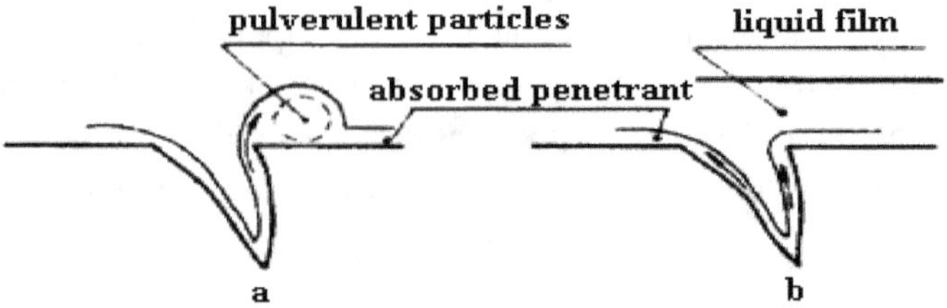

Fig. 3.30 Development of the penetrant: a. with developer containing pulverulent particles; b. with liquid developer

3.6.3 Performing determinations on GFRP materials by liquid penetrant flaw detection

As indicated above, all six steps of the liquid penetrant determination have been complied with for the samples prepared as specified.

3.6.3.1 Surface preparation

Surface preparation consisted of careful cleaning of the surfaces. One problem that occurred here was that cleaning with acetone-based products, thinners, etc., could not be done as they would damage the hardened epoxy resin surfaces, which form the matrix of the composite material. The use of water is also not a solution because of the possibility of its absorption by the composite material, especially through the areas subject to chipping. In these conditions, cleaning was carried out by wiping with dry cotton material, repeating the operation several times.

3.6.3.2 Application of penetrating liquid

After cleaning the surfaces, the penetrating liquid was applied. Red liquid penetrant was used, the container for which is shown in Figure 3.31a. Coating was done by spraying from the container, spraying perpendicularly to the coating surfaces. This is done for better infiltration of the penetrating liquid into the capillaries at the sample surface.

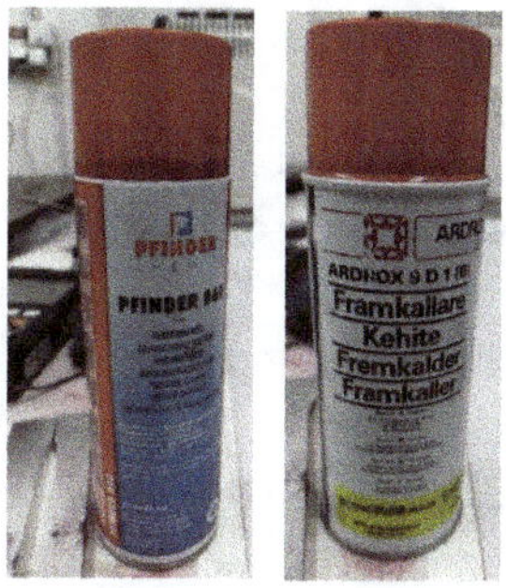

a) Red penetrating liquid **b) Developer - white**

Fig. 3.31 Consumables used

The penetrating liquid coated samples can be seen in Figure 3.32. The notation of the samples and their positioning remained the same, as shown in Figure 3.6. Considering the possibility and the ability of the composite material used to absorb the penetrating liquids, the penetrating liquid coating did not last more than 5 minutes. A longer duration may lead to false indications of propagated cracks due to uncontrolled dispersion of the penetrating liquid inside (in volume) the material. From Figure 3.32, it can be seen that along the notch, a smaller sized groove was also fabricated by previous mechanical machining in order to initiate fatigue propagated cracking more quickly and possibly in a more controlled manner. However, no fatigue propagated cracking can be observed at this point in the penetrating fluid coating.

Fig. 3.32 Liquid penetrant coating

3.6.3.3 Removing excess penetrating fluid

After coating with liquid penetrant and waiting for the liquid to penetrate for a certain time (five minutes), the third step of the determination was carried out, i.e. the removal of excess penetrant, Figure 3.33. This was carried out under a water jet in two ways: first with a water jet at normal velocity and, after removal of the vast majority of the excess penetrant, with a water jet at high velocity. Caution: in the second case, the water jet must be applied at an angle to the surface of the specimen in order to avoid water penetrating into fatigue cracks.

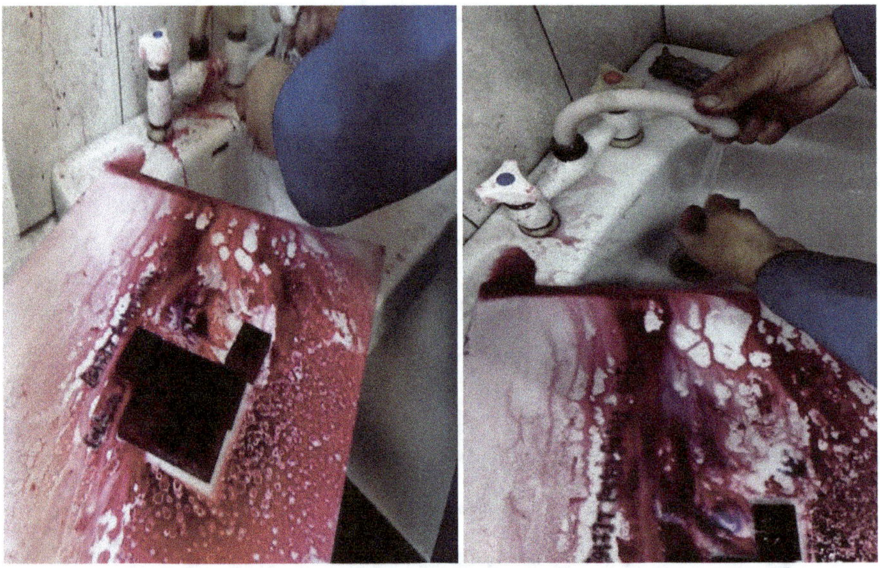

a) Normal speed; b) High speed

Fig. 3.33 Washing samples with water jet

3.6.3.4 Drying surfaces

After washing them, the samples were dried naturally in the sun on both sides. Typically, a forced-draft, directed heat source is used to dry metal parts. GFRP samples cannot be dried in this way and natural drying was used. However, this drying could not last very long because of the uncontrolled possibility of absorption of the penetrating liquid by the material, especially through the previously mechanically machined areas.

3.6.3.5 Applying the developer coating

The next step was the application of the white developing liquid on the surface of the samples, Figure 3.34. The application was made by spraying from the container shown in Figure 3.31b.

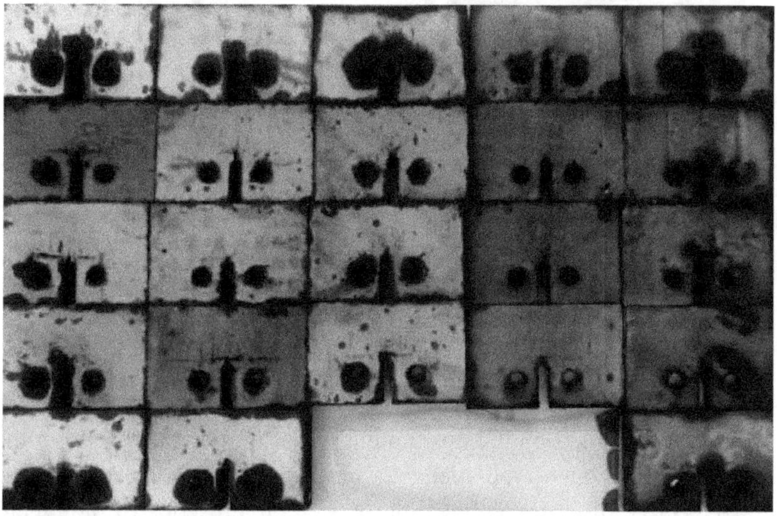

Fig. 3.34 Developer-coated samples after liquid penetrant coating and water jet washing

It is mentioned that, in order to better highlight fatigue cracks, the developing liquid was applied to both surfaces of the specimens. It is known that even for metallic specimens, which have a compact structure, the fatigue crack may have different lengths on the two surfaces. This happens because of the delay in crack propagation in one of the two surfaces. The crack front is certainly coherent, but it may be asymmetric with respect to an axis passing through the middle of the resulting fracture surface, Figure 3.35.

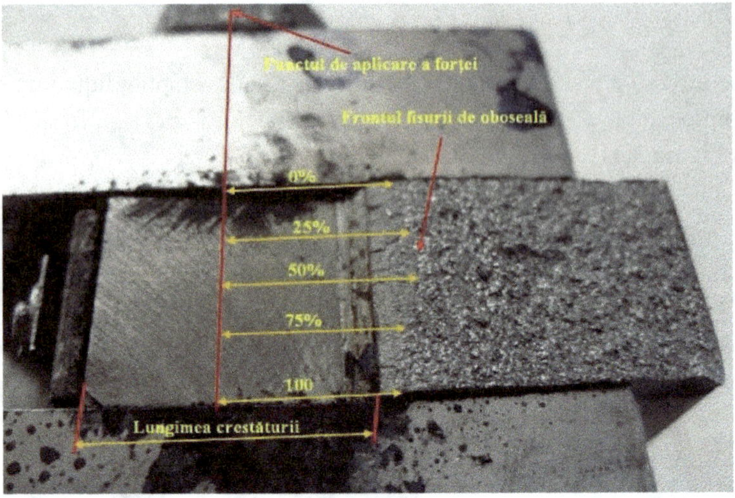

Fig. 3.35 Asymmetric appearance of crack propagated through a metal sample (Goanță V., 2006)

All the more, this phenomenon can be observed in composite specimens, where the structure is determined by the positioning of the reinforcement. From the four or five specimens of the same type available, one representative specimen was chosen from the perspective of crack propagation. A stereomicroscope of the type OPTIKA SN 255308 was used for all five specimens examined (for the figures marked a) and an optical microscope of the type Leica 5000 DMI was used for the other figures (b, c, d).

Figure 3.36 shows TR 3 [-45°/0°/+45°/90°] and shows the following:

- In all 4 pictures, there are cracks propagated by fatigue;
- In all 4 pictures, the bidirectional fabric type is highlighted;
- In Figure 3.36a, a picture was taken of the area of interest, with different magnification powers 50X, 100X and 200X, analyzing the propagated cracks in more depth;
- The length of the propagated cracks meets the requirements of the standard for determining the fracture mechanics sizes to be between 2 and 5 mm.;
- All cracks propagate along the notch formed by mechanical processing, with small deviations from the median plane, for some of them;
- In Figure 3.36b, the crack propagated parallel to the direction of the weft can be observed.

The crack lengths were measured and entered into the crack propagation calculation program as part of the main determinations.

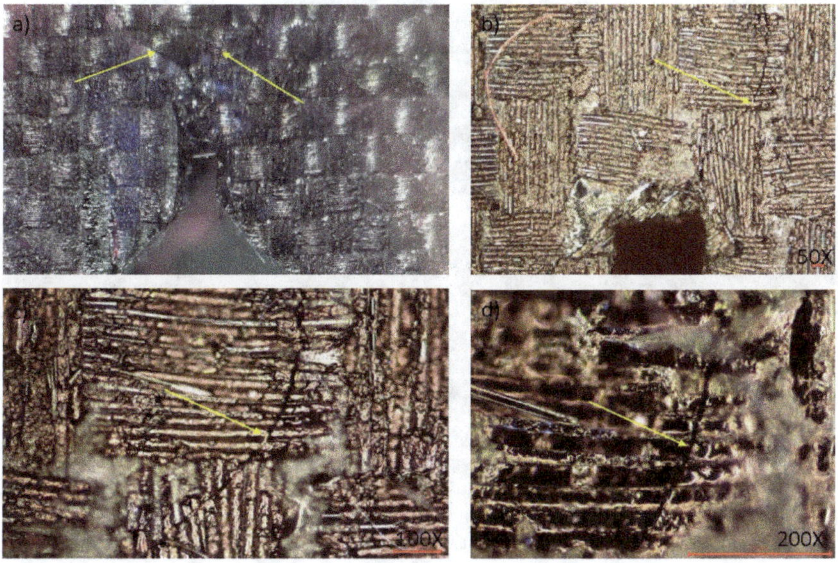

Fig. 3.36 Detail of sample TR 3 [-45°/0°/+45°/90°] – transversal cut

Figure 3.37 shows the sample LG 4 [-45°/0°/+45°/90°] and the following is noted:

- In all 4 pictures, there are cracks propagated through fatigue;
- In all 4 pictures, the bidirectional fabric type is highlighted;
- In Figure 3.37a, 2 cracks can be observed that propagated from the direction of the notch formed by mechanical processing;
- Figure 3.37d shows a break in the matrix.

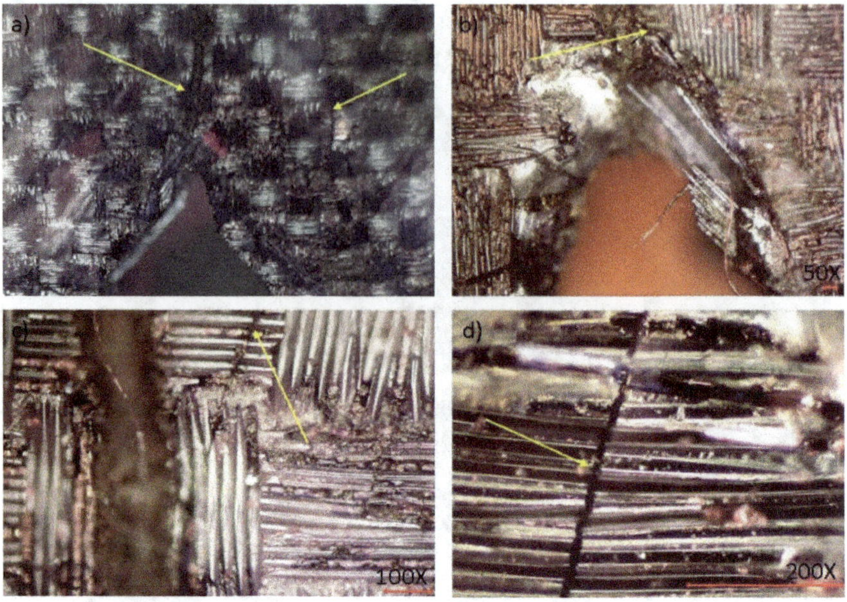

Fig. 3.37 Detail of sample LG 4 [-45°/0°/+45°/90°] – transversal cut

Figure 3.38 shows sample LG 1 [0°/90°] and shows the following:

- There are a variety of crack propagation modes;
- Figure 3.38b shows a matrix break and an initial deviation from the crack midplane;
- Figure 3.38c shows cracks in the matrix layer;
- Pronounced delaminations and voids in the resin layer are observed in Figure 3.38d.

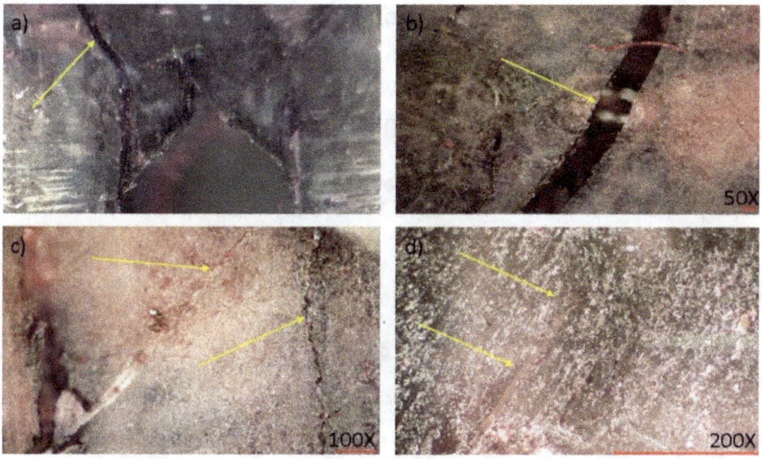

Fig .3.38 Detail of sample LG 1 [0°/90°] – longitudinal cut

Figure 3.39 shows sample TR 1 [0°/90°]-and shows the following:

- In some samples, the crack did not propagate along the mid-plane of the notch;
- Fatigue propagated cracks are observed in all previously requested samples;
- In samples where the crack did not propagate in a straight line, measuring the crack length was quite difficult;
- In Figure 3.39b and Figure 3.39c, you can see cracks in the matrix.

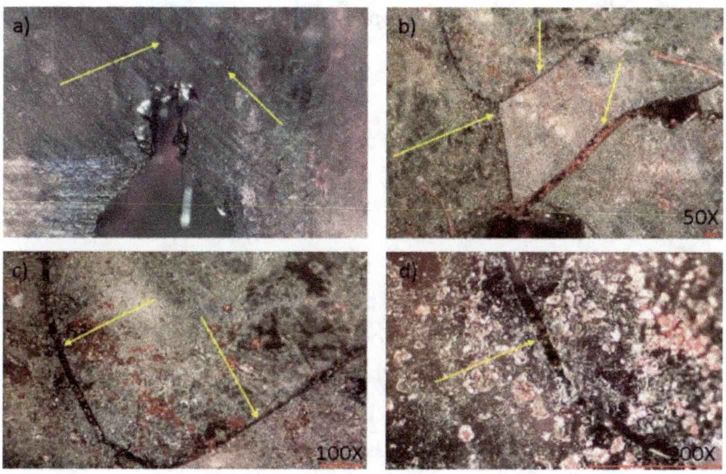

Fig. 3.39 Detail of sample TR 1 [0°/90°] – transversal cut

Figure 3.40 shows specimen 4 [0°/90°]-diagonal cut and shows the following:

- Figure 3.40b shows a crack with a considerable length but also a crack starting from the point of application of one of the forces.
- In Figure 3.40d, after matrix layer failure, fiber pull-outs and voids in the fiber-matrix interface can be seen.

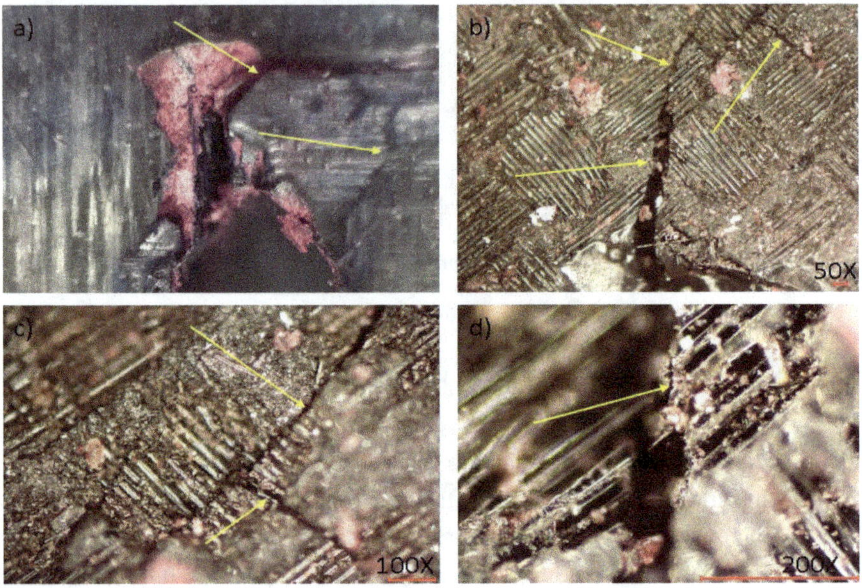

Fig. 3.40 Detail of sample 4 [0°/90°] – diagonal cut

3.7 Determinations of fracture toughness on CT samples from GFRP

3.7.1 Theoretical considerations on fracture toughness determination

In the last decade, due to its attractive characteristics such as high strength-to-weight ratio and excellent corrosion resistance, composite materials such as glass fiber reinforced polymer (GFRP) have seen considerable growth in various applications. It is necessary to understand the fracture behavior of composite materials, such as: matrix cracking, delamination, translaminar fracture, and fiber breakage, especially in aircraft structures. A failure in a structure will tend to propagate and eventually lead to the collapse of the structure. This problem is solved by testing in the area of Fracture Mechanics, which provides information about the critical stress intensity factors near the crack tip at failure. Therefore, accurate standard tests must be used to measure the fracture properties of composite materials (F.A McClintock, 1971; A.K. Head, 1969; N.E. Frost,1958; P.C. Paris, 1963). The compact tensile (CT) test is used to determine the translaminar cracking strength of GFRP composites (P.C.Paris, 1965; A.K. Head, 1969)

There is ASTM E399 (Standard Test Method for Linear-Elastic Plane-Strain Fracture Toughness of Metallic Materials) which deals with the determination of the fracture toughness of metallic materials. There is also ASTM Standard E1922 (Standard Test Method for Translaminar Fracture Toughness of Laminated and Pultruded Polymer Matrix Composite Materials) which deals with the determination of translaminar fracture toughness of polymer matrix composite materials. Each of these standards utilizes a compact tensile specimen with the specification that determinations based on ASTM E1922 must be made with great care, as the results are strictly valid only under conditions where the damage area at the notch tip is small compared to the notch length, and crack propagation occurs in the mid-plane of the notch.

3.7.2. Material and samples

The materials used are the same as those presented previously. The GFRP plates described above were used, the cutting being done under the same conditions and in the same directions, the names of the samples being:

- [-45°/0°/+45°/90°] - longitudinal (cutting direction);
- [-45°/0°/+45°/90°] – transversal (cutting direction);
- [0°/90°]– longitudinal (cutting direction);
- [0°/90°]– transversal (cutting direction);
- [0°/90°]– diagonal (cutting direction);

The geometry and overall dimensions of the sample used are those in Figure 3.41, being closer to the requirements of ASTM E1922.

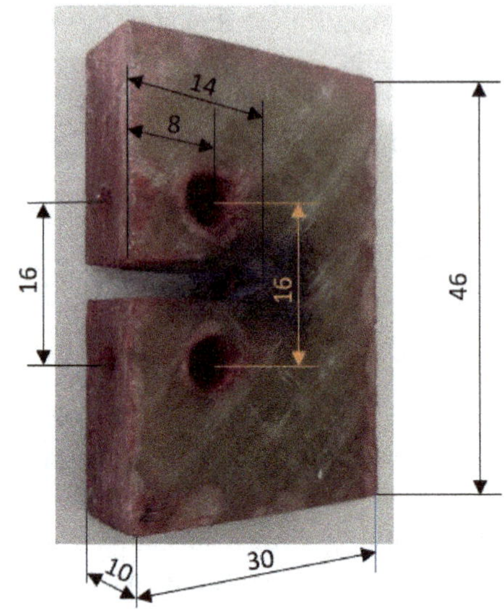

Fig. 3.41 The configuration and dimensions of the sample used for determining fracture toughness

Tensile testing of the specimens is performed on the Instron 8801 universal machine, as shown in Figure 3.7. The loading is static, with the variation of force versus displacement of the extensometer application points recorded.

3.7.3. Interpreting the force-displacement graph

The force-displacement record obtained during the test can have one of the forms shown in Figure 3.42. The assessment of the test results for the determination of the fracture toughness $\mathbf{K_{Ic}}$ consists in determining the calculated fracture toughness value $\mathbf{K_Q}$ from the force-displacement record. The calculated fracture toughness value is determined on the basis of the critical force $\mathbf{P_Q}$, which is obtained as follows:

- an OB secant is drawn with a slope 5% less than the slope of the initial portion of the force-displacement (OA) record;
- the force corresponding to the intersection of the secant OB with the recorded curve is denoted $\mathbf{P_5}$.

If the force-displacement record is of type I, Figure 3.42, i.e. all force values preceding $\mathbf{P_5}$ are smaller, then $\mathbf{P_Q} = \mathbf{P_5}$. If a maximum of the force preceding force $\mathbf{P_5}$

is recorded (Type II and III records in Figure 3.42), then the computational force value $\mathbf{P_Q}$ is taken equal to the maximum force value in the record preceding $\mathbf{P_5}$. The non-linearities occurring on the initial portion of the record, due to the placement of the displacement translator, are neglected.

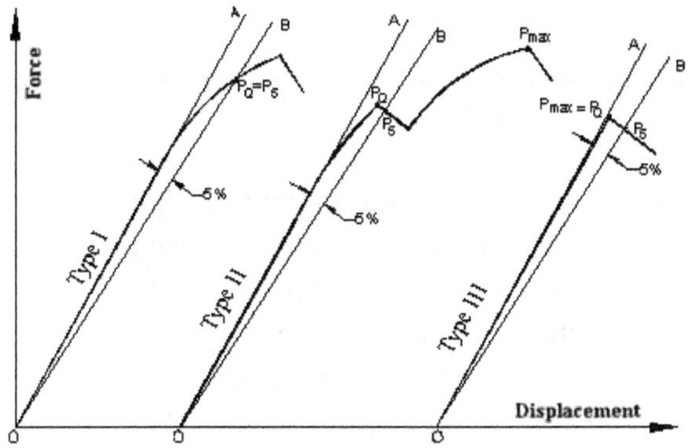

Fig. 3.42 Types of force-displacement variations under application to determine K_{Ic}

Determine the P_{max}/P_Q ratio and, if it is less than 10, calculate K_Q.

Calculated fracture toughness $\mathbf{K_Q}$ is determined from the following relationship

$$K_Q = F_Q \frac{f\left(\frac{a}{W}\right)}{BW^{\frac{1}{2}}} \tag{3.5}$$

where:

$$f\left(\frac{a}{W}\right) = 29{,}6\left(\frac{a}{W}\right)^{\frac{1}{2}} - 185{,}5\left(\frac{a}{W}\right)^{\frac{3}{2}} + 655{,}7\left(\frac{a}{W}\right)^{\frac{5}{2}} - 1017\left(\frac{a}{W}\right)^{\frac{7}{2}} + 638{,}9\left(\frac{a}{W}\right)^{\frac{9}{2}} \tag{3.6}$$

After calculating the $\mathbf{K_Q}$ value, check if:

$$B, a \geq 2{,}5\left(\frac{K_{Ic}}{\sigma_c}\right)^2 \tag{3.7}$$

If the condition (3.7) is satisfied, the critical value fracture toughness $\mathbf{K_{Ic}}$ is taken to be equal to the calculated value of the intensity factor $\mathbf{K_Q}$, thus:

$$K_{Ic} = K_Q \tag{3.8}$$

If condition (3.7) is not satisfied, the test shall be repeated on larger specimens complying with the conditions of the plane strain.

3.7.4 Experimental results obtained from tests to determine crack toughness on GFRP samples

When the test is started, the window shown in Figure 3.43 appears, in which, during the test, the force-displacement graph for the extensometer application points (crack opening displacement - COD) is to be plotted. After the stress is applied until the specimen breaks, the graph is finalized by plotting two straight lines, also visible in Figure 3.42, which must be: the first straight line is tangent to the elastic zone and the second straight line is inclined to the first one by 5%. As it can be seen in Figure 3.43, the software failed to correctly identify the straight-line tangent to the elastic zone, and hence there is an error in the estimation of the force F_Q to be considered and used in the relation (3.5) for calculating the crack toughness.

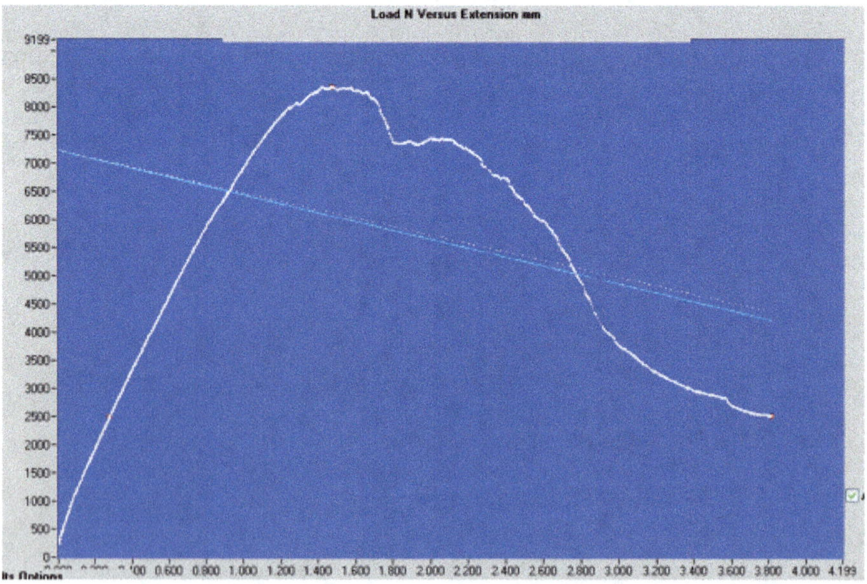

Fig. 3.43 Variation of the tension force in relation to the displacement of the force application points - for the [-45°/0°/+45°/90°] longitudinal sample

Considering the previous observation, the option of using the automatic determination of crack toughness by the "K_{Ic}" software will be abandoned, and the calculations shown in equations (3.5) and (3.6) will be performed manually, based on the data provided by the testing machine software, by processing them through the manual plotting of the two straight lines mentioned above. Under these conditions, the value obtained for the force F_Q will be much closer to reality, in accordance with one of the possibilities illustrated in Figure 3.42

Taking into consideration the above observation, we will abandon the possibility of using the automatic determination of the fracture toughness by the "K_{Ic}" software and we will make the calculations shown by relations (3.5) and (3.6), in virtue of the data received from the testing machine software, processing it by manually drawing the two lines mentioned above. Under these conditions, a value much closer to reality will be obtained for the F_Q force, in accordance with one of the possibilities shown in Figure 3.42.

Specimen LG4 [-45°/0°/+45°/90°] – longitudinal cut

As it can be seen in Figure 3.45a and in view of the previous considerations, the lines drawn by the testing machine software for this specimen are incorrectly placed. Consequently, the software has considered the value of 8.36 kN for the force F_Q with which it has calculated the fracture toughness: K_{Ic}= 34.3 MPa-m$^{1/2}$. In order to correct it, the operations described below will be performed.

In Excel, only the elastic zone is considered from the force versus displacement curve of the extensometer application points, Figure 3.44. Automatically draw the approximation line which gives the equation seen in Figure 3.44. The equation being of the type: y=mx+n, it is found that m=6573.5 is the slope of this line. As a result, the line of intersection with the entire force-displacement graph will be of equation: y= 6244.83d+622.31, where the slope of the new line is obtained as: 6244.83 = (6573.5-5/100*6573.5).

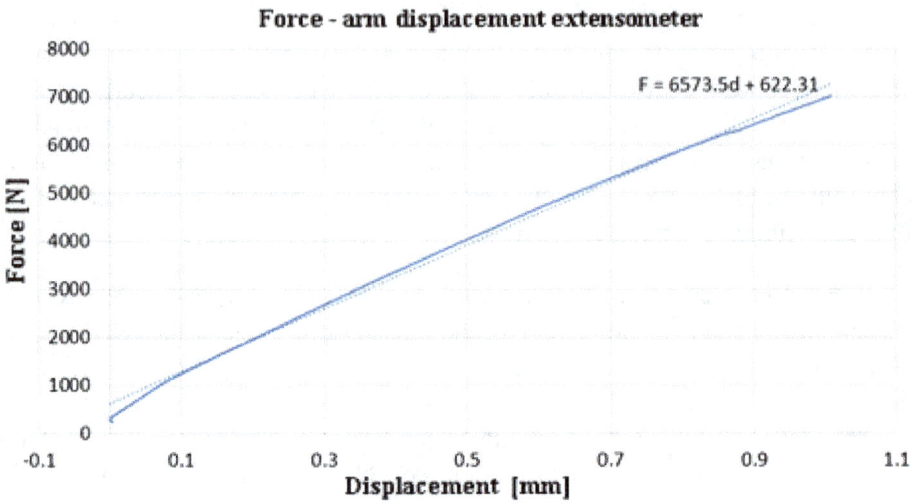

Fig. 3.44 Plotting the line of elasticity with determination of the equation -sample [-45°/0°/+45°/90°]- longitudinal cut

The data for plotting the slope of the line at 5% with reference to the elastic line are:

Y_elastic=	6753.5d+622.31
m_elastic=	6573.5
m_correct 5%=	6244.825
Y_slope5%=	6244.83d+622.31

In the complete force-displacement graph, a line is drawn that intersects the graph at the coordinate point (1.1; 7491.623), visible in Figure 3.45b. As a result, the force required to be introduced into relation (3.5) is: F_Q=7491.623 N.

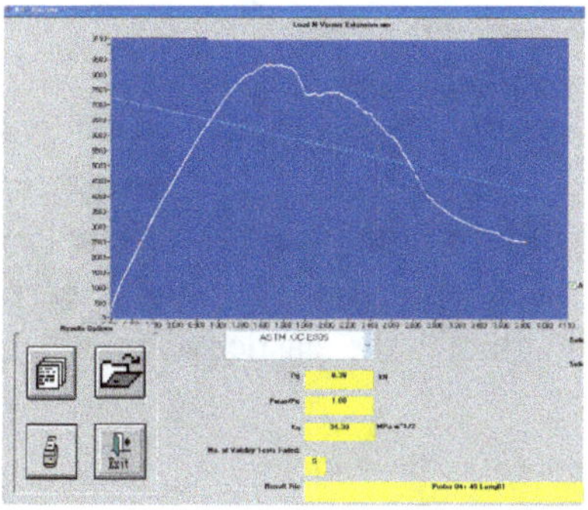

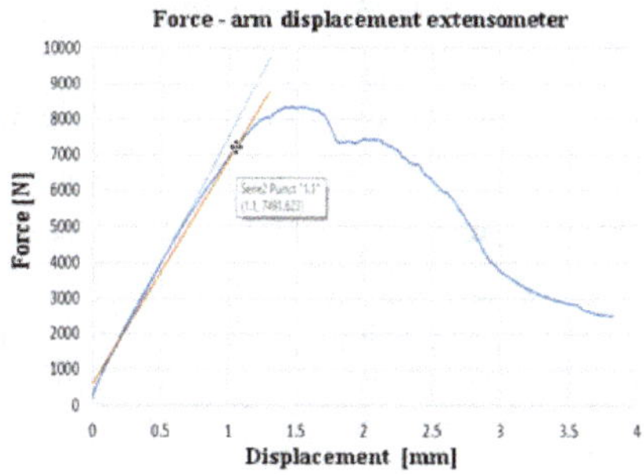

a) **image taken from the PC of the testing machine**

b) **force-displacement graph with Excel processing**

Fig. 3 45 Graphs of force variation in relation to the displacement of the extensometer application points – for the sample [-45°/0°/+45°/90°]– longitudinal cut

Some of the data required for the calculation of F(a/W) and K_{Ic} were initially entered. For the crack length, the data is taken from the file with the .DAT extension. Table 3.3 presents these data for the LG4 [-45°/0°/+45°/90°] sample cut longitudinally; the calculation of the fracture toughness being made using the relationship (3.5).

Note: the necessary transformations are made so that the fracture toughness results in [Mpa·m$^{1/2}$].

Table 3.3 Data entered for sample [-45°/0°/+45°/90°] longitudinal cut

Parameter name	Symbol	Units	Value
Force	F_Q	N	7491,623
Width	W	Mm	21,23
Thickness	B	Mm	10,34
Crack length	A	Mm	6,3
Function	f(a/w)	-	5,813
Fracture toughness	K_{Ic}	[Mpa·m$^{1/2}$]	28,909

Consequently, the value of 28.9 [Mpa·m$^{1/2}$] for the fracture toughness can be considered representative for the longitudinal cut [-45°/0°/+45°/90°] sample.

In the following, for the other samples, the calculations will be done similarly and only the final results will be presented.

Specimen TR3 [-45°/0°/+45°/90°] – transversal cut

For this sample, the curve in the elastic domain of the variation of force, F, in relation to the points of application of the extensometer arms, d, is shown in Figure 3.46. The equation of the line drawn automatically from the facilities offered by the Excel program is: F=5368.2d +408.2. It has been observed that, in this case, the variation of F-d is well approximated by a straight line in the elastic zone.

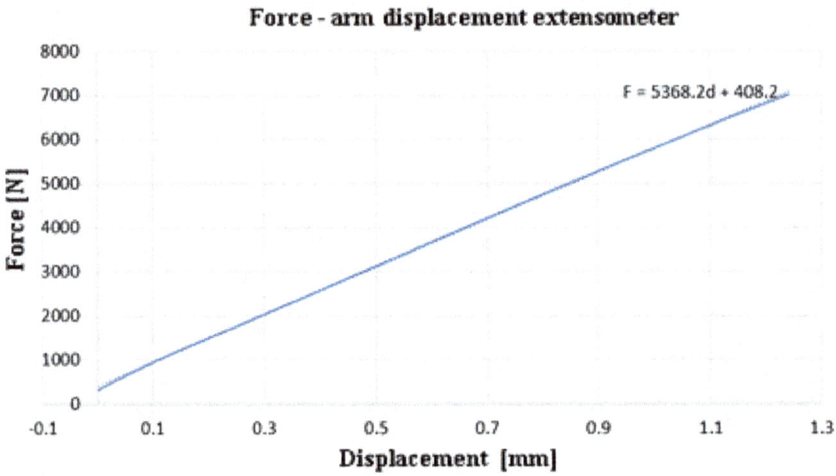

Fig. 3.46 Plotting the line of elasticity with determination of the equation -sample [-45°/0°/+45°/90°]- transversal cut

The data for plotting the slope of the line at 5% with reference to the elastic line are:

Y_elastic=	5368.2d+408.2
m_elastic=	5368.2
m_ correct 5%=	5099.79
Y_slope5%=	5099.79d+408.2

In Excel, a data table is generated and it contains: in the first column, the displacements from 0.1 to 0.1 mm are given (entered by the user); in the second column, the force values for the straight line with a slope of 5% are given, resulting from the application of the relationship: F=5099.79d+408.2:

d [mm]	Straight line 5%
0	408.2
0.1	918.179
0.2	1428.158
0.3	1938.137
0.4	2448.116
0.5	2958.095
0.6	3468.074
0.7	3978.053
0.8	4488.032
0.9	4998.011
1	5507.99
1.1	6017.969
1.2	6527.948
1.3	7037.927
1.4	7547.906
1.5	8057.885
1.6	8567.864

In Figure 3.47b, this line is drawn, which, at the intersection with the graph of force variation in relation to the displacement of the extensometer arms, provides the coordinate point (1.6;8567.864).

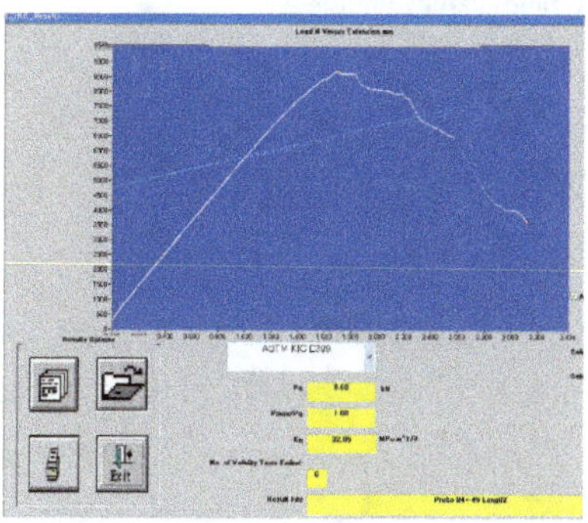

a) image taken from the PC of the testing machine

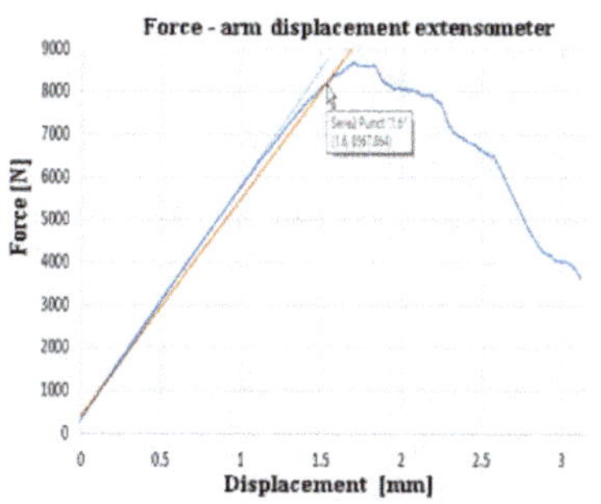

b) force-displacement graph with Excel processing

Fig.3.47 Graphs of force variation in relation to the displacement of the extensometer application points – for the sample [-45°/0°/+45°/90°]– transversal cut

The force value of 8567.864N is entered in the data Table 3.4, together with other data characteristic of the [-45°/0°/+45°/90°] transversely cut sample. The fracture toughness K_{Ic} will be automatically calculated in Excel, making some transformations so that the value results in [Mpa·$m^{1/2}$].

Table 3.4 Data entered for sample [-45°/0°/+45°/90°] transversal cut

Parameter name	Symbol	Units	Value
Force	F_Q	N	8567,864
Width	W	Mm	21,23
Thickness	B	Mm	10,34
Crack length	A	Mm	7,2
Function	f(a/w)	-	6,341
Fracture toughness	K_{Ic}	[Mpa·$m^{1/2}$]	36,058

Consequently, the value of 36,058 [Mpa·$m^{1/2}$] for the fracture toughness can be considered representative for the sample [-45°/0°/+45°/90°] transversal cut.

Specimen LG1 [0°/90°] – longitudinal cut

For this sample, the curve in the elastic domain of the variation of force, F, in relation to the points of application of the extensometer arms, d, is shown in Figure 3.48.

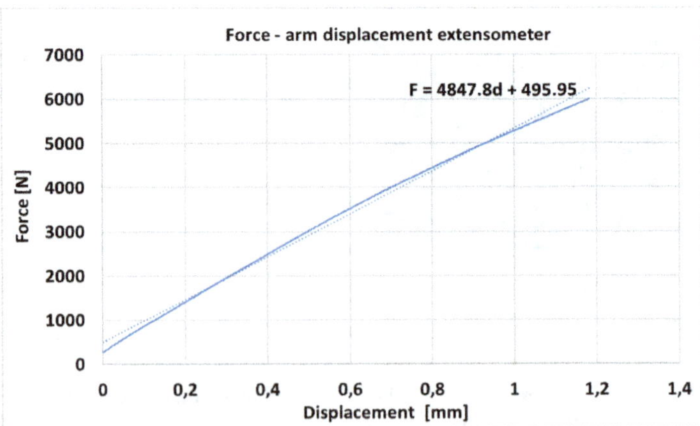

Fig.3.48 Plotting the line of elasticity with determination of the equation -sample [0°/90°]-longitudinal cut

The equation of the line drawn automatically from the facilities provided by the Excel program is: F = 4847.8d + 495.95.

The data for plotting the slope of the line at 5% with reference to the elastic line are:

Y_elastic=	4847d+495.95
m_elastic=	4847
m_correct 5%=	4604.65
Y_slope5%=	4604,65d+495.95

In Excel, a data table is generated that contains: in the first column, the displacements from 0.1 to 0.1 mm are given (entered by the user); in the second column, the force values for the line with a slope of 5% are given, resulting from the application of the relationship: F=4847d+495.95.

In Figure 3.49b, this line is drawn, which, at the intersection with the graph of force variation in relation to the displacement of the extensometer arms, provides the coordinate point (1.2;6021.53).

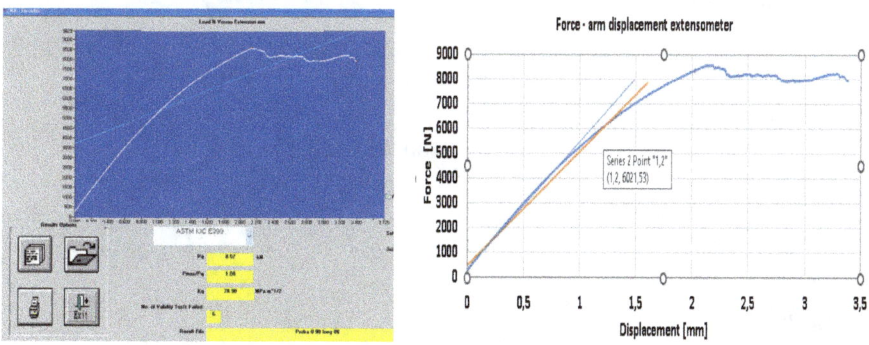

a) image taken from the PC of the testing machine

b) force-displacement graph with Excel processing

Fig. 3.49 Graphs of force variation in relation to the displacement of the extensometer application points – for the sample [0°/90°]– longitudinal cut

The force value of 6021.53N is entered in data Table 3.5, together with other characteristic data of the [0°/90°] longitudinally cut sample. The fracture toughness K_{Ic} will be automatically calculated in Excel, making some transformations so that the value results in [$Mpa \cdot m^{1/2}$].

Table 3.5 Data entered for sample [0°/90°] longitudinal cut

Parameter name	Symbol	Units	Value
Force	F_Q	N	6021,53
Width	W	Mm	21,23
Thickness	B	Mm	10,34
Crack length	A	Mm	5,6
Function	f(a/w)	-	5,506
Fracture toughness	K_{Ic}	[Mpa·m$^{1/2}$]	22,005

Consequently, the value of 22.005 [Mpa·m$^{1/2}$] for the fracture toughness can be considered representative for the longitudinally cut [0°/90°] sample.

Specimen TR1 [0°/90°] – transversal cut

For this sample, the curve in the elastic domain of the variation of force, F, in relation to the points of application of the extensometer arms, d, is shown in Figure 3.50. The equation of the line drawn automatically from the facilities provided by the Excel program is: F = 5480.7d + 553.83.

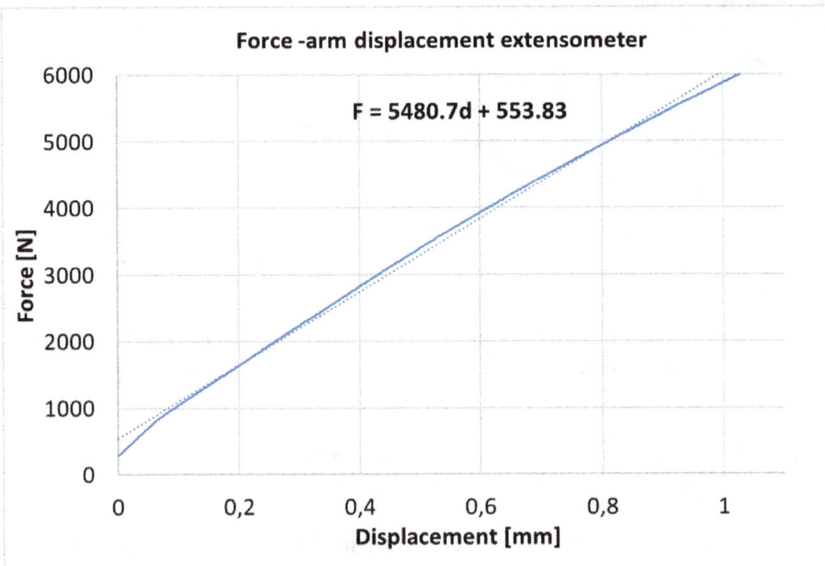

Fig. 3.50 Plotting the line of elasticity with determination of the equation -sample [0°/90°]-transversal cut

The data for plotting the slope of the line at 5% with reference to the elastic line are:

Y_elastic=	5480,7d + 553.83
m_elastic=	5480,7
m_correct 5%=	5206,665
Y_slope5%=	5206,665d+553,83

In Excel, a data table is generated that contains: in the first column, the displacements from 0.1 to 0.1 mm are given (entered by the user); in the second column, the force values for the line with a slope of 5% are given, resulting from the application of the relationship: F=5480.7d + 553.83.

In Figure 3.51b, this line is drawn, which, at the intersection with the graph of force variation in relation to the displacement of the extensometer arms, provides the coordinate point (1.1;6280.06).

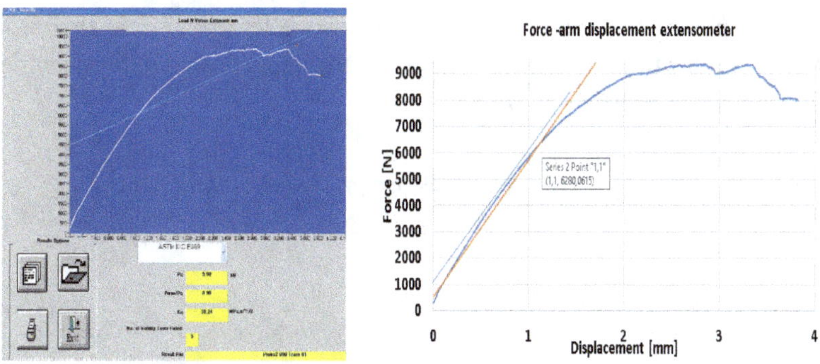

a) image taken from the PC of the testing machine

b) force-displacement graph with Excel processing

Fig. 3.51 Graphs of force variation in relation to the displacement of the extensometer application points – for the sample [0°/90°]– transversal cut

The force value of 6280.06 is entered in data Table 3.6, together with other data characteristic of the [0°/90°] transversely cut sample. The fracture toughness K_{Ic} will be automatically calculated in Excel, making some transformations so that the value results in [Mpa·m$^{1/2}$].

Table 3. 6 Data entered for sample [0°/90°] transversal cut

Parameter name	Symbol	Units	Value
Force	F_Q	N	6280,06
Width	W	Mm	21,23
Thickness	B	Mm	10,34
Crack length	A	Mm	6,95
Function	f(a/w)	-	6,181
Fracture toughness	K_{Ic}	[Mpa·m$^{1/2}$]	25,763

Consequently, the value of 25.763 [Mpa·m$^{1/2}$] for the fracture toughness can be considered representative for the [0°/90°] cross-section sample.

Specimen 4 [0°/90°] – diagonal cut

For this sample, the curve in the elastic domain of the variation of force, F, in relation to the points of application of the extensometer arms, d, is shown in Figure 3.52. The equation of the line drawn automatically from the facilities provided by the Excel program is: F = 3916.3d + 791.5

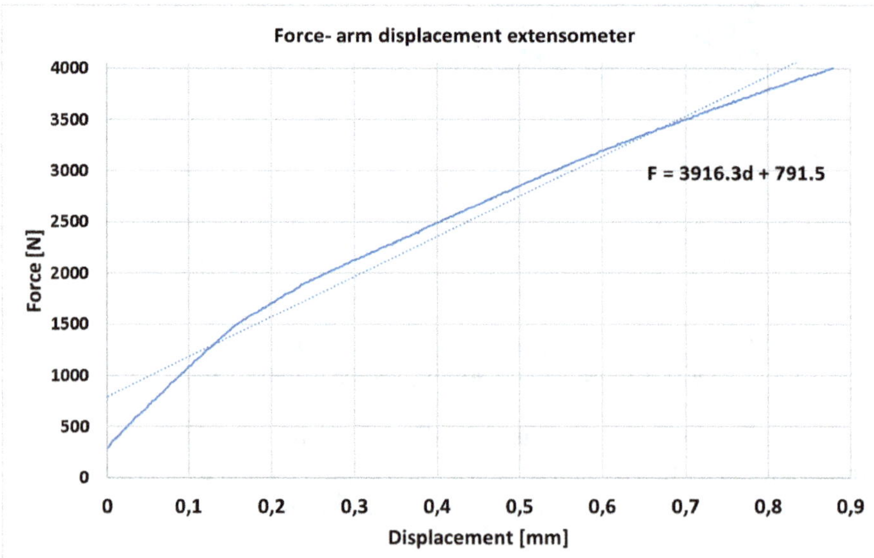

Fig. 3.52 Plotting the line of elasticity with determination of the equation – sample [0°/90°]- diagonal cut

The data for plotting the slope of the line at 5% with reference to the elastic line are:

Y_elastic=	3916.3d + 791.5
m_elastic=	3916.3
m_correct 5%=	3720.485d
Y_slope5%=	3720.485d+791.5

In Excel, a data table is generated that contains: in the first column, the displacements from 0.1 to 0.1 mm are given (entered by the user); in the second column, the force values for the line with a slope of 5% are given, resulting from the application of the relationship: F=3720.485d+791.5.

In Figure 3.53b, this line is drawn, which, at the intersection with the graph of force variation in relation to the displacement of the extensometer arms, provides the coordinate point (0.8; 3767.888).

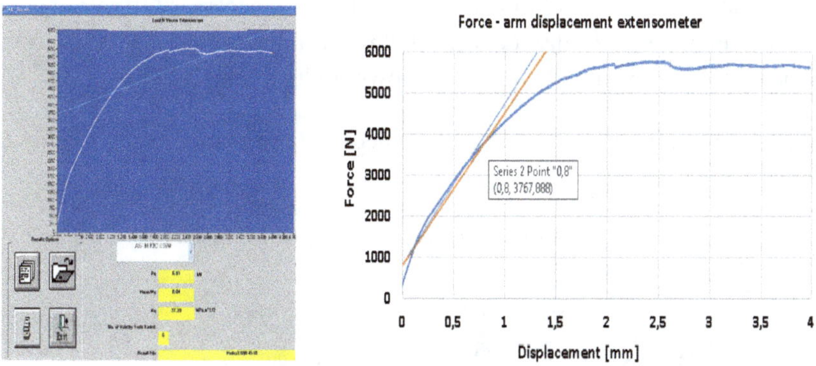

a) image taken from the PC of the testing machine **b) force-displacement graph with Excel processing**

Fig. 3.53 Graphs of force variation in relation to the displacement of the extensometer application points – for the sample [0°/90°]– diagonal cut

The force value of 3767.888 is entered in data Table 3.7, together with other characteristic data of the [0°/90°] sample cut diagonally. The fracture toughness K_{Ic} will be automatically calculated in Excel, making some transformations so that the value results in [Mpa·$m^{1/2}$].

Table 3.7 Data entered for sample [0°/90°] diagonal cut

Parameter name	Symbol	Units	Value
Force	F_Q	N	3767,888
Width	W	Mm	21,23
Thickness	B	Mm	10,34
Crack length	a	Mm	6,83
Function	f(a/w)	-	6,107
Fracture toughness	K_{Ic}	[Mpa·m$^{1/2}$]	15,274

Consequently, the value of 15,274 [Mpa·m$^{1/2}$] for the fracture toughness can be considered representative for the [0°/90°] sample cut diagonally.

Thus, it was observed that the data for force and displacement do not start exactly from zero and, probably as a result of this non-conformity, both the elasticity line and the one inclined by 5% in relation to it are drawn by the software arbitrarily. Under these conditions, our intervention was necessary, so that, by calculating and drawing the graphs in Excel, we could determine the load F_Q necessary to introduce into the calculation relationship of the fracture toughness. In the same context, all calculations were made in Excel, some of which are shown in the previous tables.

Table 3.8 summarizes the values obtained for the fracture toughness for the five test cases.

Table 3.8 Fracture toughness for GFRP samples

Nr. Crt.	Specimen name	K_{Ic} [Mpa·m$^{1/2}$]
1	Plate [-45°/0°/+45°/90°] – specimen cut longitudinally	28,909
2	Plate [-45°/0°/+45°/90°] – specimen cut transversely	36,058
3	Plate [0°/90°] – specimen cut longitudinally	22,005
4	Plate [0°/90°] – specimen cut transversely	25,763
5	Plate [0°/90°] – specimen cut diagonally	15,274

In Figure 3.54 a graph has been drawn illustrating the values obtained for the fracture toughness for the five types of samples. In this graph, the samples have been arranged according to the increasing values of K_{Ic}. It is found that the lowest value of the fracture toughness is recorded for the sample cut diagonally (neither on the weft nor on the warp) from the plate [0°/90°] and the highest value of K_{Ic} is recorded for the sample cut transversely (in the weft direction) from the plate [-45°/0°/+45°/90°].

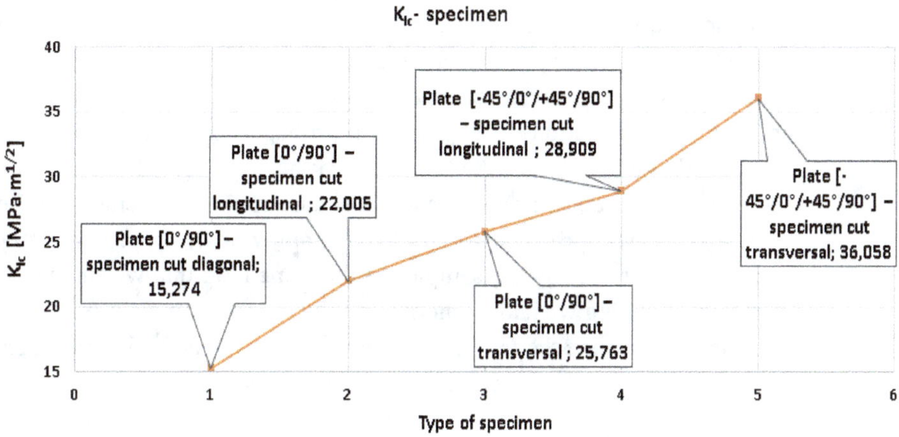

Fig. 3.54 Values obtained for K_{Ic}.

The values determined for the crack propagation speed for the five cases are not necessarily close, resulting in a certain fact: from the point of view of fracture toughness, the way the fibers are arranged in relation to the direction of stress is of great importance. For materials with fiber orientation at [-45°/0°/+45°/90°], it is most advantageous to place the fibers in such a way that the direction of normal stresses is in the weft direction, where the fracture toughness is highest. For materials with a weave at [0°/90°], it is recommended to place them in such a way that the direction of normal stresses is in the weft direction, where the fracture toughness is highest.

3.8 General conclusions to Chapter 3

1. It can be seen that the lowest value of fatigue crack growth rate is recorded for the transversely flowed specimen TR3 (in the direction of the weft) from plate [0°/90°] and the highest value of da/dN is recorded for the longitudinally flowed specimen LG4 (in the direction of the warp) from plate [0°/90°];
2. Significant difference occurs between the TR3 transversely cross-flowed sample from plate [-45°/0°/+45°/90°] (12 MP-m$^{1/2}$) compared to the TR1 transversely cross-flowed sample from plate [0°/90°] (22,28 MP-m$^{1/2}$);

3. As previously mentioned, several identical samples were made for the same plate and same direction. We expected from the outset a scatter in the results in terms of crack propagation mode in the preformed notch. This is because the composite material is not a structural material, being anisotropic with differently arranged reinforcement layers;
4. Consequently, as the defectoscopic determinations indicated, there was some variation in the direction of fatigue crack propagation for the same type of samples. We note that once the fatigue crack has propagated, the use of the specimens in subsequent determinations is easier, since the cracks in these determinations will propagate easily from the previously propagated fatigue cracks. However, among the samples of the same kind that we subjected to fatigue pre-cracking we had to choose one;
5. The tests to determine the fracture toughness showed that the lowest fracture toughness value was recorded for the 45 degree (neither weft nor warp) cut specimen from plate [0°/90°] and the highest K_{Ic} value was recorded for the cross (weft direction) cut specimen from plate [-45°/0°/+45°/90°].

Bibliography

1. Raport al comisiei către parlamentul european, consiliu, comitetul economic și social european și comitetul regiunilor https://eur-lex.europa.eu/legal-content/RO/TXT/HTML/?uri=CELEX:52021DC0950&from=EN.
2. http://free-energy-monitor.com/index.php/energy/harta_potential_eolian.
3. Raport anual privind activitatea ANRE 2021.
4. F. Blaabjerg, Ke Ma, Wind Energy Systems, Proc. of IEEE, Vol. 105, No. 11, November 2017, uploaded at 2021, DOI: 10.1109/JPROC.2017.2695485.
5. M.A.S. Shohag, E. C. Hammel, D. O. Olawale, O. I. Okoli, Damage mitigation techniques in wind turbine blades: A review. Wind Eng. 2017, 41, 185–210.
6. L. Jr. Mishnaevsky, K. Thomsen, Costs of repair of wind turbine blades: Influence of technology aspects. Wind Energy 2020, 23, 2247–2255.
7. T. Burton, D. Sharpe, N. Jenkins, E. Bossanyi, Wind energy handbook. New York: John Wiley & Sons; 2001.
8. B. Yang, D. Sun, Testing, inspecting and monitoring technologies for wind turbine blades: A survey, Renewable and Sustainable Energy Reviews 22 (2013) 515–526.
9. R. Steigmann, N. Iftimie, A. Savin, R.Sturm, Wind turbine blade composites assessment using non-contact ultrasound method. J. Clean Energy Technol. 2016, 4, 440–443.
10. http://www.caithnesswindfarms.co.uk/fullaccidents.pdf, Wind turbine accident and incident compilation.
11. Xiaowen Song, Zhitai Xing, Yan Jia, Xiaojuan Song, Chang Cai, Yinan Zhang, Zekun Wang, Jicai Guo and Qingan Li, Review on the Damage and Fault Diagnosis of Wind Turbine Blades in the Germination Stage, Energies 2022, 15, 7492. https://doi.org/10.3390/en15207492.
12. https://www.windfarmbop.com/wind-turbine-blade-damage-detection-systems.
13. Abdel-Hamid I. Mourad, Abdulla Almomani, Irfan Ahmad Sheikh, Ammar H. Elsheikh, Failure analysis of gas and wind turbine blades: A review, Engineering Failure Analysis 146 (2023) 107107.
14. L. Mishnaevsky Jr, Root causes and mechanisms of failure of wind turbine blades: overview, Materials 15 (9) (2022) 2959.
15. Al-Khudairi, Ghasemnejad, To improve failure resistance in joint design of composite wind turbine blade materials, Renew. Energy 81 (2015) 936–951.
16. B.F Soerensen, Joergensen, Improved design of large wind turbine blade of fibre composites based on studies of scale effects, 2004.

17.Abdul Ghani Olabi, Tabbi Wilberforce, Khaled Elsaid, Enas Taha Sayed, Tareq Salameh, Mohammad Ali Abdelkareem and Ahmad Baroutaji, A Review on Failure Modes ofWind Turbine Components, Energies 2021, 14, 5241. https://doi.org/10.3390/en14175241].
18.S. Ataya and Ahmed MMZ (2013) Damages of wind turbine blade trailing edge: Forms, location, and root causes. Engineering Failure Analysis 35: 480–488.
19.Yajie Yua, Hui Caoa, Xingyu Yana, Tao Wanga, Shuzhi Sam Ge, Defect identification of wind turbine blades based on defect semantic features with transfer feature extractor, Neurocomputing, 376 (2020) 1-9 https://doi.org/10.1016/j.neucom.2019.09.071.
20.X. Chen, et al., Preliminary failure investigation of a 52.3 m glass/epoxy composite wind turbine blade, Eng. Fail. Anal. 44 (2014) 345–350.
21.https://www.lamontagne.fr/saint-agnant-de-versillat-23300/faits-divers/decouvrez-enimagesleoliennecasseepresdelasouterrainecreuse_14055928/##CLICK_URL##.
22.Caithness Windfarm Information Forum. Summary of Wind Turbine Accident Data to 31 March 2021.
23.M. Ideno, K. Seki, Study on Improvement of Performance of wind power generation system and lightning damage. In Proceedings of the 28th International Conference on Lightning Protection, Kanazawa, Japan, 18–22 September 2006; pp. 1585–1589.]
24.https://www.windpowerengineering.com/blades-key-turbine-component-no-treat-way/ .
25.http://guillermo-rein.blogspot.com/2014/01/wind-turbines-on-fire.html.
26.https://www.windpowerengineering.com/the-true-cost-of-wind-turbine-fires-and-protection/.
27.H.F.Zhou , H.Y.Dou , L.Z.Qin , Y.Chen , Y.Q.Ni , J.M.Ko, A review of full-scale structural testing of wind turbine Blades, RenewableandSustainableEnergyReviews33(2014)177–187.
28.K. Branner, A. Ghadirian, Database about blade faults. DTU Wind Energy. (DTU Wind Energy E; No. 0067). 2014.
29.P.D.Bârsănescu et.al., Tensometrie electrică rezistivă aplicată la materiale compozite., Editura Tehnopress Iaşi 2004.
30.F. Pierron, Material Testing 2.0: A brief review, Strain. 2023; 59:e12434, https://doi.org/10.1111/str.12434.
31.M. Grédiac, F. Pierron, Int. J. Numer. Methods Eng. 1998, 41(2), 293; M. Grédiac, F. Pierron, Y. Surrel, Exp. Mech. 1999, 39(2), 142.
32.M. Grédiac, F. Pierron, Int. J. Numer. Methods Eng. 2004, 59(10), 1287.
33.L. Jiang, B. Guo, H. Xie, Acta Mech. Sin. 2015, 31(2), 173.

34.T. He, L. Liu, A. Makeev, Compos. Struct. 2018, 184, 337.
35.M. Grédiac, F. Pierron, A. Vautrin, Compos. Sci. Technol. 1994, 51(3), 409.
36.F. Pierron, M. Grediac, Compos. Part A Appl. Sci. Manufac. 2000, 31(4), 309.
37.H. Chalal, F. Meraghni, F. Pierron, M. Grédiac, Compos. Part A: Appl. Sci. Manufact. 2004, 35(7–8), 841.
38.H. Chalal, S. Avril, F. Pierron, F. Meraghni, Compos. Part A: Appl. Sci. Manufact. 2006, 37(2), 315.
39.F. Pierron, G. Vert, R. Burguete, S. Avril, R. Rotinat, M. R. Wisnom, Strain 2007, 43(3), 250
40.C. Gogu, W. Yin, R. Haftka, P. Ifju, J. Molimard, R. Le Riche, A. Vautrin, Exp. Mech. 2013, 53(4), 635.
41.G. Seon, A. Makeev, J. D. Schaefer, B. Justusson, Appl. Sci. 2019, 9(13), 2647.
42.C. Kim, J.-H. Kim, M.-G. Lee, Composites, Part B 2020, 200, 108338.
43.R. Moulart, S. Avril, F. Pierron, Compos. Part A: Appl. Sci. Manufact. 2006, 37(2), 326.
44.F. Pierron, S. Zhavoronok, M. Grédiac, Int. J. Solids Struct. 2000, 37(32), 4437.
45.R. Gras, H. Leclerc, F. Hild, S. Roux, J. Schneider, Int. J. Solids Struct. 2015, 55(0), 2.
46.R. Moulart, F. Pierron, S. R. Hallett, M. R. Wisnom, Exp. Mech. 2011, 51(4), 509.
47.F. Pierron, H. Zhu, C. Siviour, Philos. Trans. R. Soc., A 2014, 372(2023), 20130195.
48.L. Fletcher, J. Van-Blitterswyk, F. Pierron, J. Dynam. Behav. Mater. 2019, 5(1), 65.
49.S. Parry, L. Fletcher, F. Pierron, J. Dynam. Behav. Mater. 2021, 7(1), 127.
50.Van Blitterswyk, L. Fletcher, F. Pierron, J. Dynam. Behav. Mater. 2018, 4, 543.
51.J. Van Blitterswyk, L. Fletcher, F. Pierron, J. Dynam. Behav. Mater. 2020, 6(3), 373.
52.https://wyomingtestfixtures.com/ products/compression/ modified-astm-d-695-compression-test-fixture-boeing-bss-7260/.
53.J. Ribrant and L. M. Bertling, Survey of failures in wind power systems with focus on Swedish wind power plants during 1997–2005 IEEE Trans. Energy Conversion 22(1) pp 167-173
54.O. T. Thomsen, Sandwich materials for wind turbine blades—present and future Journal of Sandwich Structures and Materials 11(1) pp 7-26.
55.P. Veers et al, Trends in the design, manufacture and evaluation of wind turbine blades. Wind Energy, 6, pp. 245–259, 2003.
56.V. Goanta, A. Hadar and B. Leitoiu, Experimental Procedure Designed to Determine the Elastic Characteristics of Fiber-Reinforced Polymeric Composite Materials Materiale Plastice 47(4) pp 450-456, 2010.

57.R. Steigmann, A. Savin, V. Goanta and P. D. Barsanescu, Determination of mechanical properties of some glass fiber reinforced plastics suitable to Wind Turbine Blade construction IOP Conf. Series: Materials Science and Engineering, 2016, 147 012140 doi:10.1088/1757-899X/147/1/012140.
58.M. D. Stanciu, H. T Drăghicescu, I. C. Roşca, Mechanical Properties of GFRPs Exposed to Tensile, Compression and Tensile–Tensile Cyclic Tests. Polymers 2021, 13, 898.
59.M. M. Z. Ahmed, B. Alzahrani, N. Jouini, M. M. Hessien, and S. Ataya, The Role of Orientation and Temperature on the Mechanical Properties of a 20 Years Old Wind Turbine Blade GFR Composite. Polymers 2021, 13, 1144.
60.J Abenojar, M Pantoja, MA Martinez and JC del Real, Aging by moisture and/or temperature of epoxy/SiC composites: Thermal and mechanical properties, J Compos. Mater.49(24), pp 2963-2975, 2015.
61.J Chaichanawong, C Thongchuea and S Areerat, Effect of moisture on the mechanical properties of glass fiber reinforced polyamide composites, Adv Powder Technol 27, pp 898-902, 2016.
62.HS Choi, K J Ahn, J-D Nam and HJ Chun, Hygroscopic aspects of epoxy/carbon fiber composite laminates in aircraft environments, Compos Part A 32(5), pp 709-720, 2001.
63.M. Cordin, T. Bechtold, T. Pham, Effect of fibre orientation on the mechanical properties of polypropylene–lyocell composites. Cellulose 2018, 25, 7197–7210.
64.C.Cerbu, Aspecte privind degradarea caracteristicilor elastice şi mecanice de încovoiere ale materialelor compozite sticlã / polimer din cauza absorbtiei de umiditate, revista MATERIALE PLASTICE 44 Nr. 2 , 2007, Rezultatele unor cercetări expermentale
65.J. M. Corum, R. L. Battiste, M. B. Ruggles and W. Ren W, Durability – bassed design criteria for a chopped-glass-fiber automotive structural composite Composite Science and Technology nr. 61 pp 1083-1085.
66.F. Pomies, L. A. Carlson and J. W. Gillespie, Marine environmental effects on polymer matrix composites Composite Materials: Fatique and Fracture, nr. 5 ASTM STP 1230 Martin R H, Ed., American Society for Testing and Materials, Philadelphia, pp. 283-286.
67.L B Manfredi, De Santis and A. Vázquez, Influence of the addition of montmorillonite to the matrix of unidirectional glass fibre/epoxy composites on their mechanical and water absorption properties Composites: Part A 39 pp1726-1731, 2008.

68.K. Imielinska and L. Guillaumat , The effect of water immersion ageing on low-velocity impact behaviour of woven aramid-glass fibre/epoxy composites Composites Science and Technology 64 pp 2271-2278.
69.F.A Ramirez, L.A Carlsson and B. A. Acha, Evaluation of environmental effects on polymer matrix composites by micromechanical and macromechanical tests 13th European Conference on Composites Materials (13-ECCM), Stockholm, Sweden, June 2-5, 2008.
70.P. Sampath Rao, Dr. M. Manzoor Hussain, R. Kishore, Moisture Absorption Evolution of Gfrp Laminates Subjected To Different Environmental Conditions; IOSR Journal of Mechanical and Civil Engineering (IOSRJMCE) ISSN: 2278-1684 Volume 2, Issue 5 (Sep-Oct 2012), PP 33-38.
71.M Eftekhari and A Fatemi, Tensile behavior of thermoplastic composites including temperature, moisture, and hygrothermal effects, Polym Test 51, pp 151-164, 2016.
72.S Eslami, F Taheri-Behrooz and F Taheri, Effects of Aging Temperature on Moisture Absorption of Perforated GFRP, Adv Mater Sci Eng 2012, 7pp, Article ID 303014, doi:10.1155/2012/303014, 2012.
73.Y.C.Lin and X. Chen, Moisture sorption-desorption-resorption characteristics and its effect on the mechanical behavior of the epoxy system. Polymer 2005; 46: 11994–12003.
74.X. Han, A. D. Crocombe, S. N. R. Anwar, et al, The effect of a hot-wet environment on adhesively bonded joints under a sustained load. J Adhes 2014; 90: 420–436.
75.Behzad Ahmed Zai, M. K. Park, H. S. Choi, Hassan Mehboob and Rashid Ali, Effect of moisture absorption on damping and dynamic stiffness of carbon fiber/epoxy composites, Journal of Mechanical Science and Technology 23 (2009) 2998~3004.
76.ASTM D3039/D3039M-08 Standard Test Method for Tensile Properties of Polymer Matrix Composite Materials.
77.C. Morăraş, D. E. Husaru, Structural changes of GFRP composite material after immersion in liquid nitrogen and normal water, IOP Conf. Series: Materials Science and Engineering 1262 (2022) 012055, doi:10.1088/1757-899X/1262/1/012055.
78.A. Hadăr, Probleme locale la materiale compozite, Teză de doctorat, U.P.B., 1997.
79.E. Alămoreanu, C. Negruţ, G. Jiga, Calculul structurilor din materiale compozite, U.P.B, 1993.
80.P. D. Barsanescu, F. Mocanu, L. Bejan, C. Bitca , Tensometrie electrica rezistiva aplicata la materiale compozite, Ed. Tehnopress, Iasi, 2005, ISBN 973-702-100-2.
81.ASTM D3518/D3518M-18, Standard Test Method for In-Plane Shear Response of Polymer Matrix Composite Materials by Tensile Test of a ±45° Laminate.

82. ASTM D 5379/D 5379M – 98, Standard Test Method for Shear Properties of Composite Materials by the V-Notched Beam Method; V-Notched Beam Test Coupon Schematic.
83. C. Morăraș, C. A. Tugui, R. Steigmann, P. D. Barsanescu, B. Leitoiu, V. Goanta, Mechanical Testing of GFRP Composite Materials Used in Wind Turbine Blades Construction, Advanced Materials Research, 2020, ISSN: 1662-8985, Vol. 1157, pp 142-148.
84. R. R. Conant and E. M. Odom, An improved Iosipescu shear test fixture. J. Comp. Tech. Res. 17 (1); 50-55, 1995
85. ASTM D3410/D3410M, Standard Test Method for Compressive Properties of Polymer Matrix Composite Materials with Unsupported Gage Section by Shear Loading.
86. V. Goanță, Mecanica ruperii, Ed. Tehnopress, Iași, 2006.
87. Gh. Amza, A. Mihai, V. Popovici, G. Jiga, A.Raduta and V. Goanta, Editura Academia Oamenilor de Stiinta din Romania, 2011, ISBN 978-606-8371-47-4
88. P.C. Paris, G.C. Sih, Stress analysis of cracks. Fracture Toughness Testing and Its Applications, ASTM STP 381: 30-83, Philadelphia: ASTM, 1965.
89. F.A McClintock, Plasticity aspects of fracture. In Fracture: An Advanced Treatise, Vol.3: 47-225, Academic Press, New York, 1971.
90. A.K. Head, Physics and Strength of Plasticity, M.I.T. Press, 1969.
91. N.E. Frost, D.S. Dugdale, The propagation of fatigue cracks in test specimens. J Mechan Phys Solids; 6:92–110, 1958.
92. P.C. Paris, F. Erdogan, F. A critical analysis of crack propagation Laws. Transaction ASME, Journal of Basic Engineering, vol. 85, no. 4, p. 528-533, 1963.

www.ingramcontent.com/pod-product-compliance
Lightning Source LLC
LaVergne TN
LVHW022058200926
848481LV00003B/967
* 9 7 9 8 8 9 9 6 6 0 9 0 0 *